How Math Can Save Your Life

How **Math** Can
Save Your Life

JAMES D. STEIN

WILEY

John Wiley & Sons, Inc.

Published by John Wiley & Sons, Inc., Hoboken, New Jersey

Published simultaneously in Canada

For general information about our other products and services, please contact our Customer Care Department within the United States at (800) 762–2974, outside the United States at (317) 572–3993 or fax (317) 572–4002.

Wiley also publishes its books in a variety of electronic formats. Some content that appears in print may not be available in electronic books. For more information about Wiley products, visit our web site at www.wiley.com.

Library of Congress Cataloging-in-Publication Data:

Stein, James D., date.
 How math can save your life / James D. Stein.
   p. cm.
 Includes index.
 ISBN 978-0-470-43775-9 (cloth)
  1. Mathematics—Miscellanea. I. Title.
 QA99.S735 2010
 510—dc22

2009028776

Printed in the United States of America
10 9 8 7 6 5 4 3 2 1

For Maxine and James, my first arithmetic teachers

CONTENTS

13   How Math Can Improve Society   205

How much is a human life worth in dollars? • When should legal cases be settled out of court? • At what point does military spending become unnecessary?

14   How Math Can Save the World   215

Do extraterrestrial aliens exist? • How can we prevent nuclear war and a major asteroid impact? • When is the world going to end?

PREFACE

My performance in high school English courses was somewhat less than stellar, partly because I enjoyed reading science fiction a lot more than I liked to read Mark Twain or William Shakespeare. I always felt that science fiction was the most creative form of literature, and Isaac Asimov was one of its most imaginative authors.

He may not have rivaled Shakespeare in the characters or dialogue department, but he had *ideas*, and ideas are the heart and soul of science fiction. In 1958, the year I graduated from high school, Asimov's story "The Feeling of Power" appeared in print for the first time. I read it a couple of years later when I was in college and coincidentally had a summer job as a computer programmer, working on a machine approximately the size of a refrigerator whose input and output consisted of punched paper tape.

Asimov's story was set in a distant future, where everyone had pocket calculators that did all of the arithmetic, but nobody understood the rules and ideas on which arithmetic was based. We're not quite there yet, but we're approaching it at warp speed. As I got older, I noticed the decline in my students' arithmetic abilities, but it came to a head a few years ago when a young woman came to my office to ask me a question. She was taking a course in what is euphemistically called College

Algebra, which is really an amalgam of Algebra I and II as given in countless high schools. Several comments had led me to believe that the students in the class didn't understand percentages, so I had given a short quiz—for the details, see chapter 6. As the young woman and I were reviewing the quiz in my office after the exam, we came to a problem that required the student to compute 10 percent of a number.

"Try to do it without the calculator," I suggested.

She concentrated for a few seconds and became visibly upset. "I can't," she replied.

After that incident, I began to watch students in my class as they took tests. I deliberately design all of my tests so that a calculator is not needed; I'm testing how well the students can use the ideas presented in the course, not how well they can use a calculator. I can solve every single problem on every exam I give without even resorting to pencil-and-paper arithmetic, such as would generally be required to multiply two two-digit numbers or add up a column of figures. I noticed that the typical student was spending in the vicinity of 20 percent of the exam time punching numbers into a calculator. What the hell was going on?

What had happened was that the presence of calculators had caused arithmetical skills to atrophy, much as Asimov had predicted. More important, though, was something that Asimov touched on in his story but didn't emphasize in the conclusion. Here are the last few lines of the story: "Nine times seven, thought Shuman with deep satisfaction, is sixty-three, and I don't need a computer to tell me so. The computer is in my own head. And it was amazing the feeling of power that gave him."[1]

Almost all math teachers will tell you that the power of arithmetic is not the ability to multiply nine times seven, but the knowledge of the problems that could be solved by multiplication. Of course, that philosophy was behind the original rush to stick a calculator in the hands of every schoolchild as soon as he or she could push the buttons. Arithmetic had become the red-headed stepchild of mathematics education. The thought was that if we just got past the grunt work of tedious arithmetic, we could fast forward to the beauty and power of higher mathematics.

Unfortunately, we lost sight of the fact that there is a whole lot of beauty and power in arithmetic. Although most people can *do* arithmetic, few really understand and appreciate its scope. The feeling of power alluded to in the last line of Asimov's story comes nowadays not with the ability to calculate, but with the ability to use the powerful and beautiful tool that is arithmetic. Arithmetic can greatly improve the quality—and the quantity—of your life. It can improve the organizations and the societies of which you are a part. And yes, it can even help save the world.

In writing this book, I was tremendously fortunate to have help from several people. There are a few chapters on money and finance, which constitute an important model of arithmetic, and the chapters benefited considerably from my consultation with Merrick Sterling, the retired executive vice president of Portfolio Risk Management Group at the Union Bank of California. Rick retired at a sufficiently young age so that he could pursue his early love of mathematics. As a result, he acquired a master's degree and has exchanged the corner office in his bank having an exquisite view and perks for a single desk in a room shared by several part-time instructors. Talk about upward mobility! Sherry Skipper-Spurgeon, whom I met during a textbook adoption conference in Sacramento, is the hardest-working elementary and middle-school teacher I have ever encountered, and I would unhesitatingly sign on to any project whatsoever for the opportunity to work with her. She has worked on numerous state and national conferences on mathematics education in elementary schools and is knowledgeable about not only the programs in education but also the behind-the-scene politics. Robert Mena, the chair of my department, is extremely well-versed in many areas of mathematics in which I am deficient and is also a terrific teacher, which is a rare quality in an administrator. Walk into his office and the first thing you see is a wall of photographs of students who have received A's from him. A number of students have even received five A's, a tribute not only to his popularity as a teacher but to the variety of courses he teaches.

My career as an author would probably have been confined to blogging were it not for my agent, Jodie Rhodes, who once confided to me that she had sold a book after it had received more than two hundred rejections! That's tenacity rivaling, or even exceeding, that of the legendary king of Scotland Robert Bruce. I'm trying hard not to break that record.

I have also been tremendously lucky to have Stephen Power as the editor of this book. Writing a trade book in mathematics is a touchy task, especially for an academic, and Stephen deftly steered me between the Scylla of unsupported personal opinion (of which I have lots) and the Charybdis of a severe case of Irving-the-Explainer syndrome, in which teachers too often indulge. Even better, he did so with humor and instant feedback. Waiting for an editor to get back to you with comments is as nerve-racking as waiting for the results of an exam on which you have no idea how you did. If, as Woody Allen says, 80 percent of life is showing up, it's nice to work with someone who believes, as I do, that the other 20 percent is showing up promptly.

Finally, I would like to thank my wife, Linda, not only for the work she has put into proofreading this book, but also for the joy she has brought to so many aspects of my life. Marriage is a special kind of arithmetic, in which $1 + 1 = 1$.

What Math Can Do for You

We can get a good idea of how education has changed in the United States by taking a look inside a little red schoolhouse in the heartland of America a little more than a century ago.

Salina, Kansas, 1895

There's a very good chance that you are not reading these words in Salina, Kansas (current population approximately 50,000), and you're certainly not reading them in 1895. There's also a very good chance that the typical twenty-first-century American couldn't come close to passing the arithmetic section of the 1895 Salina eighth-grade exit exam. In case you're skeptical, here it is.[1]

Arithmetic (Time, 1.25 hours)

1. Name and define the Fundamental Rules of Arithmetic.

2. A wagon box is 2 ft. deep, 10 feet long, and 3 ft. wide. How many bushels of wheat will it hold?

3. If a load of wheat weighs 3,942 lbs., what is it worth at 50 cts. per bu., deducting 1,050 lbs. for tare?

4. District No. 33 has a valuation of $35,000. What is the necessary levy to carry on a school seven months at $50 per month, and have $104 for incidentals?

5. Find cost of 6,720 lbs. coal at $6.00 per ton.

6. Find the interest of $512.60 for 8 months and 18 days at 7 percent.

7. What is the cost of 40 boards 12 inches wide and 16 ft. long at $20.00 per in?

8. Find bank discount on $300 for 90 days (no grace) at 10 percent.

9. What is the cost of a square farm at $15 per acre, the distance around which is 640 rods?

10. Write a Bank Check, a Promissory Note, and a Receipt.

If I were to let you use a calculator, allow you to skip questions 1 and 10, and tell you some of the fundamental constants needed for this exam, such as the volume of a bushel of wheat (which is needed on question 2), you might still have a rough time. Yet Salina schoolchildren were supposed to be able to pass this exam without a calculator—and they had only an hour and fifteen minutes to do it.

I haven't reprinted the other sections of the exam, but this part of the exam is worth looking at because it reveals the philosophy of nineteenth-century education: prepare citizens to be productive members of society. That doesn't seem to be the goal of education anymore—at least, it's certainly not the goal of mathematics education after the basics of arithmetic have been learned. The world today is vastly more complicated than it was in Salina, Kansas, in 1895, but mathematics can play a huge role in helping to prepare citizens to be productive members of society. Regrettably, that's not happening—and it's not so hard to make it happen.

How much math do you need to be a productive citizen, to enrich your life and the groups of which you are a part? Amazingly enough, sixth-grade arithmetic will take you an awfully long way if you just use it right, and you can go further with only a few extra tools that are easy to pick up. You don't need algebra, geometry, trigonometry, or calculus.

I've been teaching college math for more than forty years, and I've worked with programs at both the primary and the secondary levels. I have yet to find a good explanation for why the math education establishment insists on stuffing algebra down everyone's throat, starting in about seventh grade. After all, who really needs algebra? Certainly, anyone planning a career in the sciences or engineering does, and it's useful in the investment arena, but that's about it. Algebra is mandatory on the high school exit exam of many states, despite overwhelming evidence that outside of the people who really need algebra (the groups mentioned previously), almost nobody needs algebra or ever uses it once they put down their pencils at the SAT. People certainly didn't bother teaching it in Salina in 1895. Salina was a rural community, most students ended up working on farms or possibly in town, and there were lots of chores to do and no point in learning something that was virtually useless for most people. We have a lot more time now, because we don't have to get up at five in the morning to milk the cows and we don't have to go right into the workforce once we finish eighth grade. You'd think we'd use the extra time to good advantage, to enable our high school graduates to get a lot more out of life. Isn't that the purpose of education?

This is a book about how the math you already know can help you get a lot more out of life from the money you spend, from your job, from your education, and even from your love life. That's the purpose of mathematics. I wish I could enable everyone to understand the beauty and power inherent in much of what is called higher mathematics, but I've been teaching long enough to know that it's not going to happen. As with any area, such as piano, the further you go in the subject the more difficult it becomes. Most piano teachers know that people who take up the piano will never play all three movements of Beethoven's *Moonlight* Sonata, but they also know that anyone can learn to play a simple melody with enough proficiency to derive pleasure from the activity. It's the same with math, except that its simple melodies, properly played, can enrich both the individual and society.

You already have more than enough technique to learn how to play and profit from a surprisingly large repertoire of mathematics, so let's get started.

1

The Most Valuable Chapter You Will Ever Read

Are service contracts for electronics and appliances
just a scam?

• • •

How likely are you to win at roulette?

• • •

Is it worth going to college?

What constitutes value? On a philosophical level, I'm not sure; what's valuable for one person may not be for others. The most philosophically valuable thing I've ever learned is that bad times are always followed by good times and vice versa, but that may simply be a lesson specific to yours truly. On the other hand, if this lesson helps you, that's value added to this chapter. And if this chapter helps you financially, even better—because there is one universal common denominator of value that everyone accepts: money.

That's why this chapter is valuable, because I'm going to discuss a few basic concepts that will be worth tens of thousands—maybe even hundreds of thousands—of dollars to you. So let's get started.

Service Contracts: This Is Worth Thousands of Dollars

A penny saved is *still* a penny earned, but nowadays you can't even slip a penny into a parking meter—so let me make this book a worthwhile investment by saving you a few thousand dollars. The next time you go to buy an appliance and the salesperson offers you a service contract, *don't even consider purchasing it*. A simple table and a little sixth-grade math should convince you.

Suppose you are interested in buying a refrigerator. A basic model costs in the vicinity of $400, and you'll be offered the opportunity to buy a service contract for around $100. If anything happens to the refrigerator during the first three years, the store will send a repairman to your apartment to fix it. The salesperson will try to convince you that it's cheap insurance in case anything goes wrong, but it's not. Let's figure out why. Here is a table of how frequently various appliances need to be repaired. I found this table by typing "refrigerator repair rates" into a search engine; it's the 2006 product reliability survey from Consumer Reports National Research Center.[1] It's very easy to read: the top line tells you that 43 percent of laptop computers need to be repaired in the first three years after they are purchased.

Repair Rates for Products Three to Four Years Old	

Product	Repair Rate (Percentage of Products Needing Repair)
Laptop computer	43
Refrigerator: side-by-side, with icemaker and dispenser	37
Rider mower	32
Lawn tractor	31
Desktop computer	31
Washing machine (front-loading)	29
Self-propelled mower	28
Vacuum cleaner (canister)	23
Washing machine (top-loading)	22
Dishwasher	21
Refrigerator: top-and-bottom freezer, with icemaker	20
Range (gas)	20
Wall oven (electric)	19
Push mower (gas)	18
Cooktop (gas)	17
Microwave oven (over-the-range)	17
Clothes dryer	15
Camcorder (digital)	13
Vacuum cleaner (upright)	13
Refrigerator: top-and-bottom freezer, no icemaker	12
Range (electric)	11
Cooktop (electric)	11
Digital camera	10
TV: 30- to 36-inch direct view	8
TV: 25- to 27-inch direct view	6

Use this chart, do some sixth-grade arithmetic, and you can save thousands of dollars during the course of a lifetime. For instance, with the refrigerator service contract, a refrigerator with a top-and-bottom freezer and no icemaker needs to be repaired in the first three years approximately 12 percent of the time; that's about one time in eight. So if you were to buy eight refrigerators and eight service contracts, the cost of the service contracts would be 8 × $100 = $800. Yet you'd need to make only a single repair call, on average, which would cost you $200. So, if you had to buy eight refrigerators, you'd save $800 − $200 = $600 by *not* buying the service contracts: an average saving of $600/8 = $75 per refrigerator. Admittedly, you're not going to buy eight refrigerators—at least, not all at once. Even if you buy fewer than eight refrigerators over the course of a lifetime, you'll probably buy a hundred or so items listed in the table. Play the averages, and just like the casinos in Las Vegas, you'll show a big profit in the long run.

You can save a considerable amount of money by using the chart. There are basically two ways to do it. The first is to do the computation as I did above, estimating the cost of a service call (I always figure $200—that's $100 to get the repairman to show up and $100 for parts). The other is a highly conservative approach, in which you figure that if something goes wrong, you've bought a lemon, and you'll have to replace the appliance. If the cost of the service contract is more than the average replacement cost, purchasing a service contract is a sucker play.

For instance, suppose you buy a microwave oven for $300. The chart says this appliance breaks down 17 percent of the time—one in six. To compute the average replacement cost, simply multiply $300 by 17/100 (or 1/6 for simplicity)—the answer is about $50. If the service contract costs $50 or more, they're ripping you off big-time. Incidentally, note that a side-by-side refrigerator with icemaker and dispenser will break down three times as often as the basic model. How can you buy something that breaks down 37 percent of the time in a three-year period? I'd save myself the aggravation and do things the old-fashioned way, by pouring water into ice trays.

Finally, notice that TVs almost *never* break down. I had a 25-inch model I bought in the mid-eighties that lasted seventeen years. Admittedly, I did have to replace the picture tube once. Digital cameras are pretty reliable, too.

The long-term average resulting from a course of action is called the expected value of that action. In my opinion, expected value is the single most bottom-line useful idea in mathematics, and I intend to devote a lot of time to exploring what you can do with it. In deciding whether to purchase the refrigerator service contract, we looked at the expected value of two actions. The first, buying the contract, has an expected value of $-\$100$; the minus sign occurs because it is natural to think of expected value in terms of how it affects *your* bottom line, and in this case your bottom line shows a loss of $100. The second, passing it up, has an expected value of $-\$25$; remember, if you bought eight refrigerators, only one would need a repair costing $200, and $200/8 = \$25$. In many situations, we are confronted with a choice between alternatives that can be resolved by an expected-value calculation. Over the course of a lifetime, such calculations are worth a minimum of tens of thousands of dollars to you—and, as you'll see, they can be worth hundreds of thousands of dollars, or more, to you. This type of cost-effective mathematical projection can be worth millions of dollars to small organizations and billions to large ones, such as nations. It can even be used in preventing catastrophes that threaten all of humanity. That's why this type of math is valuable.

Averages: The Most Important Concept in Mathematics

Now you know my opinion, but I'm not the only math teacher who believes this: averages play a significant role in all of the basic mathematical subjects and in many of the advanced ones. You just saw a simple example of an average regarding service contracts. Averages play a significant role in our everyday use of and exposure to mathematics. Simply scanning through a few sections of today's paper, I found references to the average household income, the average per-screen revenue of current motion pictures, the scoring averages of various basketball players, the average age of individuals when they first became president, and on and on.

So, what is an average? When one has a collection of numbers, such as the income of each household in America, one simply adds up all of those numbers and divides by the number of numbers. In short, an average is the sum of all of the data divided by the number of pieces of data.

Why are averages so important? Because they convey a lot of information about the past (what the average is), and because they are a good indicator of the future. This leads us to the law of averages.

The Law of Averages

The law of averages is not really a law but is more of a reasonably substantiated belief that future averages will be roughly the same as past averages. The law of averages sometimes leads people to arrive at erroneous conclusions, such as the well-known fallacy that if a coin has come up heads on ten consecutive flips, it is more likely to come up tails on the next flip in order to "get back to the average." There are actually two possibilities here: the coin is a fair coin that really does come up tails as often as it does heads (in the long run), in which case the coin is just as likely to come up heads as tails on the next flip; or the flips are somehow rigged and the coin comes up heads much more often than tails. If somebody asks me which way a coin will land that has come up heads ten consecutive times, I'll bet on heads the next time—for all I know, it's a two-headed coin.

Risk-Reward Ratios and Playing the Percentages

The phrases *risk-reward ratio* and *playing the percentages* are so much a part of the common vocabulary that we have a good intuitive idea of what they mean. The risk-reward ratio is an estimate of the size of the gain compared with the size of the loss, and playing the percentages means to select the alternative that has the most likely chance of occurring.

In common usage, however, these phrases are used qualitatively, rather than quantitatively. Flu shots are advised for the elderly because the risk associated with getting the flu is great compared with the reward of not getting it; that is, the risk-reward ratio of not getting a flu shot is high, even though we may not be able to see exactly how to quantify it. Similarly, on third down and seven, a football team will usually pass the ball because it is the percentage play: a pass is more likely than a run to pick up seven yards. There are two types of percentages: those that arise from mathematical models, such as flipping a fair coin, and those that arise from the compilation of data, such as the percentage of times a pass succeeds on third down and seven. When we flip a fair coin, we need not assume that in the long run, half of the flips will land heads and the other half tails, because that's what is meant by "a fair coin." If, however, we find out that 60 percent of the time, a pass succeeds on third down and seven, we will assume that in the long run this will continue to be the case, because we have no reason to believe otherwise unless the structure of football undergoes a radical change.

How, and When, to Compute Expected Value

The utility of the concept of expected value is that it incorporates both risk-reward ratios and playing the percentages in a simple calculation that gives an excellent quantitative estimate of the long-term average payoff from a given decision.[2] Expected value is used to compute the long-term average result of an event that has different possible outcomes. The casinos of the world are erected on a foundation of expected value, and roulette wheels provide an easy way to compute an example of expected value. A roulette wheel has 36 numbers (1 through 36), half of which are red and half of which are black. In the United

States, the wheel also has 0 and 00, which are green. If you bet $10 on red and a red number comes up, you win $10; otherwise, you lose your $10. To compute the expected value of your bet, suppose you spin the wheel so that the numbers come up in accordance with the laws of chance. One way to do this is to spin the wheel 38 times; each of the 38 numbers—1 through 36, 0, and 00—will come up once (that's what I mean by having the numbers come up in accordance with the laws of chance). Red numbers account for 18 of the 38, so when these come up, you will win $10, a total of $18 \times \$10 = \$180$. You will lose the other 20 bets, a total of $20 \times \$10 = \$200$. That means that you lose $20 in 38 spins of the wheel, an average loss of a little more than $.52. Your expected value from each spin of the wheel is thus $-\$.52$, and the casinos and all of those neon lights are built on your contribution and those of your fellow gamblers.

Expected value is frequently expressed as a percentage. In the preceding example, you have an average loss of about $.52 on a wager of $10. Because $.52 is 5.2% of $10, we sometimes describe a bet on red as having an expected value of -5.2%. This enables us to compute the expected loss for bets of any size. Casinos know what the expected value of a bet on red is, and they can review their videotapes to see whether the actual expected value approximates the computed expected value. If this is not the case, maybe the wheel needs rebalancing, or some sort of skullduggery is taking place.

Expected value can be used only in situations where the probabilities and associated rewards can be quantified with some accuracy, but there are a lot of these. Many of the errands I perform require me to drive some distance; that's one of the drawbacks of living in Los Angeles. Often, I have two ways to get there: freeways or surface streets. Freeways are faster most of the time, but every so often there's an event (an accident or a car chase) that causes lengthy delays. Surface streets are slower, but one almost never encounters an event that turns a surface street into a parking lot, as can happen on the freeways.

Nonetheless, like most Angelenos, I have made an expected-value calculation: given a choice, I take the freeway because on average I save time by doing so. It is not always necessary to perform expected-value calculations; simple observation and experience give you a good estimate of what's happening, which is why most Angelenos take the freeway. You don't have to perform the calculation for the roulette wheel, either; just go to Vegas, make a bunch of bets, and watch your bankroll dwindle over the long run.

Insurance: This Is Worth Tens of Thousands of Dollars

There's a lot of money in the gaming industry, but it pales in comparison with another trillion-dollar industry that is also built on expected value. I'm talking about the insurance industry, which makes its profits in approximately the same way as the gaming industry. Every time you buy an insurance policy, you are placing a bet that you "win" if something happens that enables you to collect insurance, and that you "lose" if no such event occurs. The insurance company has computed the average value of paying off on such an event (think of a car accident) and makes certain that it charges you a large enough premium that it will show a profit, which will make your expected value a negative one.

Nonetheless, this is a game that you simply have to play. If you are a driver, you are required to carry insurance, and there are all sorts of insurance policies (life, health, home) that it is advisable to purchase, even though your expected value is negative—because you simply cannot afford the cost of a disaster. Despite that, there is a correct way to play the insurance game, and doing this is generally worth tens of thousands of dollars (maybe more) over the course of a lifetime.

Let's consider what happens when you buy an auto insurance policy, which many people do every six months. My insurance company offers me a choice of a $100 deductible policy for $300 or a $500 deductible policy for $220. If I buy the $100 deductible policy and I get into an accident, I get two estimates for the repair bill and go to the mechanic who gives the cheaper estimate (this is standard operating procedure for insurance companies). The insurance company sends me a check for the amount of the repair less $100. If I had bought the $500 deductible policy, the company would have sent me a check for the amount of the repair less $500. It's cheaper to buy the $500 deductible policy than the $100 deductible policy, because if I get in an accident, the insurance company will send me $400 less than I would receive if I'd bought the $100 deductible policy.

An expected-value calculation using your own driving record is a good way to decide which option to choose. I've been driving fifty years and bought a hundred six-month policies. During that period, I've had three accidents. One was my fault— I wasn't paying attention. The other two both occurred during a three-day period in 1983: in each case, I was *not even moving* and a car rammed into me and totaled my vehicle. I am getting older, however, and am probably not as good a driver as my record shows, so I estimate that having one accident every five years is probably a little more accurate than having three in fifty years. This means that if I buy ten policies (two every year for five years) and choose the $100 deductible, rather than the $500 dollar deductible, I'll save $80 the nine times out of ten that I don't have an accident and lose $400 the one time that I do. So, by buying the $100 deductible, I save an average of $32, because (9 × $80 − $400)/10 = $32. It actually figures to be somewhat more than that for two reasons. I think that the estimate of one accident every five years is a little conservative, but, more important, if I have an accident that doesn't have to be reported (for instance, if I accidentally back up too far and hit the wall of my garage), I just might pay for the repair myself, because I know my insurance rates will skyrocket once I file a claim.

This calculation occurs countless times, as the deductible option is presented to you every time you buy health insurance or any kind of property insurance as well—and you and your family will purchase an extraordinary amount of insurance during the course of a lifetime. For some people, the savings from making the correct decisions will be in the hundreds of thousands of dollars, but for everyone it's at least in the tens of thousands—unless you're a Luddite who has rejected modern technology.

Because a crucial factor of the calculation is an estimate of the likelihood of certain events occurring, it's important to have a plan to figure this out. When purchasing auto insurance, I use my own driving record, but if you are just starting out, a reasonable approach is to use the accident statistics of people in a group similar to yours. If you are a twenty-five-year-old woman, look for accident statistics for women between twenty and thirty years old; numerous Web sites exist that contain this or similar information. If you are considering buying earthquake insurance, find out something about the frequency of earthquakes where you live. If you live in an area that has never experienced an earthquake, why would you want to buy earthquake insurance?

Let's Take a Break

You might be a little weary from all of these calculations. Fortunately, today is the day that you will go to a taping of your favorite game show. Like many game shows, it has a preliminary round in which the contestant wins some money. The host then tries to persuade the contestant to risk that money in an attempt to win even more. Incredibly, you have been selected from the studio audience to be a contestant on such a game show, you have successfully managed to answer who was buried in Grant's tomb, and you have won $100,000. The host congratulates you on the depth of your knowledge, and a curtain is drawn back onstage, revealing three doors. The host informs you that behind one of these doors is a check for $1,000,000, and behind the other two is a year's supply of the sponsor's product, which happens to be toothpaste. The host tells you that in addition to the $100,000 that you have already won, you get to pick a door, and you will receive whatever lies behind that door.

Three has always been your lucky number, so you go with door three. The host walks over to door three, hesitates—and turns the handle on door two. Tubes of toothpaste cascade all over the stage. The host, now knee-deep in toothpaste, turns and says, "Have I got a deal for you! You can either keep the $100,000 and whatever lies behind door three, or you can give me back the $100,000 and take what lies behind door one instead." Well, what do you do?

I give this question to every class in which I teach probability and ask the students what they would do. To a man (or a woman), they keep the $100,000 and whatever lies behind door three. After all, a bird (or $100,000) in the hand is not something most people are comfortable letting get away.

The correct answer to this problem actually involves a consideration of external factors. For instance, if you have a child who needs a critical operation that costs exactly $100,000 and this is your only way of getting the money, of course you would keep the $100,000. This $100,000 is worth far more to you than the $1,000,000 you might receive in addition; economists have devised a concept called *marginal utility* to describe the fact that each extra dollar beyond the $100,000 needed for the operation has significantly less value to you than the dollars that make up the $100,000 for the operation.

Let's say, however, that you regard all dollars as having equal value and, having been placed in a game situation, feel that you are obliged to play the game to earn the most dollars in the long run. In other words, when situations such as this are presented to you, you want to make the play that gives you the greatest expected value. In this case, you should relinquish the $100,000 (albeit with regret) and take what lies behind door one—because the probability that the big prize lies behind door one is twice as great as the probability that it lies behind door three!

The first time most people encounter a situation like this, they see it as highly counterintuitive. How can it be twice as likely to be behind one door as another? Isn't it equally likely to be behind either door? Yes, but the tricky point here (occasionally, tricky points really do show up in math problems) is that you are not being asked to choose between door three and door one, you are asked to choose between door three and *the other two doors*. And it just happens that you have seen the toothpaste behind one of the other two doors. To make this a little clearer, suppose that there were a thousand doors rather than three doors, and only one of them contained a $1,000,000 check. As before, the host opens all of the doors except door three (your choice) and door one, and (this time up to his neck in toothpaste) he asks you if you want to switch. Your chance of guessing the correct door was originally 1 in 1,000, and nothing has happened to change those odds: there are 999 chances out of 1,000 that the million-dollar check is behind door one.

You can now see that in the original three-door problem, there is one chance in three that the million-dollar check lies behind your choice of door three, and two chances in three that it lies behind door one. If you stick with your original choice of door three, thinking of the toothpaste as valueless, you have two out of three chances to win $100,000 and one out of three chances to win $1,100,000, for an average win of a little more than $433,000—so $433,000 is the expected value of choosing door three. If you switch doors and pick door one, you will have one chance to win $0 (ouch) but two chances to win $1,000,000, for an average win of a few hundred short of $667,000—so $667,000 is the expected value of choosing door one.

I mentioned earlier that external considerations have to be taken into account. If you are married, switch doors and give up the $100,000, and emerge with nothing but toothpaste to show for your efforts, be prepared to listen to your spouse bring it up until the end of time.[3]

Going to College: A Decision Worth Hundreds of Thousands of Dollars

So far, we've looked at a couple of very ordinary events: buying a refrigerator and selecting an insurance policy. Now let's look at an extraordinary event: deciding whether to go to college. Although many of us go to college, the use of the word *extraordinary* is justified by the dictionary, for going to college is a one-time experience for most of us and is highly exceptional or unusual within the context of our own lives.

Back in the early 1990s, I worked on a project that involved high school teachers. One of them taught math at a high school in the San Fernando Valley and told me that he had tried to persuade one of his better students to go to college. At the last moment, the student told the teacher that he had been offered a good job in the construction industry and had decided to take that instead.

Many of the readers of this book will have faced this or a similar decision: Should I take my B.A. and get a job, or should I go to graduate school, med school, or law school? It is one of the most financially important decisions you will ever make, and there are lots of factors to take into account. It will cost money to go to college, and you may not complete it. It will take you out of the job market for several years. As against that, college graduates make considerably more than high school graduates do. What's the right thing to do?

Almost invariably, the right thing is to seek more schooling. Yes, lots of people will tell you this, but here we will do the math. In 2004, a high school graduate earned an average of about $28,000 a year, whereas a college graduate earned about $51,000 per year.[4] Even if you assume you have only a fifty-fifty chance of graduating from a public college and it costs you $50,000 to attend school for five years and graduate (the time needed by a typical student where I teach), let's look at what it's worth to you. If you are eighteen years old with a high school degree and planning on working until you are sixty-five (that's forty-seven years), the cost to you (compared with the high school graduate who goes straight into the job market) of failing to graduate after five years in college is $50,000 plus five years of earning $28,000 a year, for a total of $190,000. If, however, you graduate after five years of college, compared to the high school graduate who went straight to work, you will have lost the five years of earning $28,000 a year and the $50,000 tuition, but you will gain $23,000 per year for the forty-two years you will be in the workforce. That's a net gain of $776,000. If you were to flip a coin (analogous to the fifty-fifty chance of graduating from college) and if the coin lands heads you win $776,000, and tails you lose $190,000, your expected value is $293,000. This computation is highly conservative: the college graduation rate is generally much higher than 50 percent. If your chances of graduating are 75 percent—three out of four—you rate to win $776,000 three times and lose $190,000 once, for an average gain of $(3 \times \$776,000 - 1 \times \$190,000)/4 = \$534,500$! (It may be somewhat self-serving of me to make this remark, but my guess is that if you are reading this book, your chances of graduating from college are considerably better than fifty-fifty.) If you do the same calculation for the decision as to whether to pursue an advanced degree, the results are similar.

One Long Season

A friend of mine once had a conversation with a sports gambler who made a successful living betting the Big Three: baseball, football, and basketball. Each of these three sports has a season, and even though they overlap slightly, essentially the year consists of a baseball season, a football season, and a basketball season. The gambler told my friend that even though he liked to show a profit at the end of each season, he recognized that you win some and you lose some. The key was to regard life as one long season—you're in it to show a profit over the long haul.

The same is true with playing the percentages. Certain situations will recur, such as buying auto insurance or service contracts, and it is easy to see that the law of averages will work for you in this type of situation. Other things, however, such as deciding to go to college, are essentially one-shot affairs: although people do drop out of school and return thirty years later to pick up the sheepskin, most people who drop out for several years never come back. Nonetheless, every time you play the percentages in the long season of life, you are giving yourself the best chance of showing a profit, and over that long season this is the best strategy.

2

How Math Can Help You Understand Sports Strategy

Why could Bart Simpson probably beat you at rock, paper, scissors?

• • •

What are "pure" and "mixed" strategies?

• • •

Is a pass play or a run play more likely to make a first down?

Many of the important problems we encounter in life involve competition. Sometimes we are competing to poke our head out above the crowd, such as when we apply for a job or appear on *American Idol*. Often, though, it's just us against a single opponent—although that single opponent may be an aggregation sometimes referred to as "management" or "your parents." One-on-one conflict situations were studied extensively in the first half of the twentieth century, and an important discipline emerged: game theory.

Rock, Paper, Scissors

Many important aspects of game theory can be explained by analyzing the classic game of rock, paper, scissors—a game that, curiously enough, seems to have evolved in several different cultures. For those unfamiliar with the game, on the count of three each of the two players chooses one of the three objects by extending his hand in one of three configurations. A clenched fist represents a rock, a flat hand with the palm down represents paper, and a fist with the second and third fingers extended to make a V represents scissors. If both players choose the same object, the game is a tie. Otherwise, the winner is determined according to the following rules:

Rock breaks (defeats) scissors.
Scissors cuts (defeats) paper.
Paper covers (defeats) rock.

This game is often played several times to determine a winner: two children faced with an unpleasant chore such as washing the dishes might play rock, paper, scissors, with the first person to win three times getting to avoid the chore.

To analyze the game, let's imagine that you are forced to play against a computer that has a complete record of the thousands of games you have previously played. If you have a tendency to choose one of the objects rather than the others, the computer will ruthlessly exploit this tendency. For instance, let's suppose your history shows that you choose rock 38 percent of the time, scissors 32 percent of the time, and paper 30 percent of the time. The computer will choose paper every time, and in 100 games you will lose 38, win 32, and tie 30, for a net loss of 6. The way to prevent the computer from exploiting such a tendency is to avoid showing a preference for choosing one object, which can be done by picking each of the three objects one-third of the time.

If, however, you're playing against a perfect computer, there is another trap you must avoid. Not only must you choose each object one-third of the time, you must avoid falling into a pattern, or the computer will pick up on it and capitalize. If you were to select the three objects in a predetermined pattern, such as rock-paper-scissors-rock-paper-scissors-rock-paper-scissors, the computer would detect this and adopt the obvious countermeasure, because it would know precisely what you were going to choose. Even if you were to reveal the slightest hint of a pattern, such as choosing rock 38 percent of the time after you have chosen two consecutive scissors, the computer would pick up on it and exploit it. Therefore, you have to choose each object one-third of the time and must do so randomly, so that there is no pattern to exploit. You might do something like this: roll a six-sided die (hiding the result from the computer), and choose rock if the die shows a 1 or a 2, scissors if the die shows a 3 or a 4, and paper if the die shows a 5 or a 6. Assuming the throws of the die are perfectly random, you will choose each object one-third of the time with no apparent pattern, and even a perfect computer cannot beat you.

Yet there is a downside to selecting this particular strategy. If you happen to be playing against Bart Simpson, arguably the word's dumbest rock-paper-scissors player, who chooses rock every single time (while thinking, Good old rock. Nothing beats rock.), you will not win. Unlike Lisa Simpson, who knows that Bart always chooses rock and plays accordingly, when playing Bart you will win one-third of the time (when you choose paper), lose one-third of the time (when you choose scissors), and tie one-third of the time (when you choose rock). Anyone who has ever played any sort of a game, whether a physical game such as football or an intellectual one such as poker, will tell you that it is far more dangerous to underestimate your opponent than it is to overestimate him. Thus, game theory is devised under the assumption that you are playing against an intelligent opponent who is capable of capitalizing on any error you might make.

Rock, paper, scissors is an example of what is called a 3×3 game—each of the two players has a choice of three different strategies. Early books on game theory were written during the cold war, when the Russians were red and the Americans true-blue, and the two opponents were usually denoted *red* and *blue*. Curiously, the game was usually analyzed from the standpoint of red, a tradition to which we have adhered. In order to describe the game mathematically, the result of each possible choice was placed in the form of a matrix.

		Blue	
	Rock	Paper	Scissors
Red			
Rock	0	−1	1
Paper	1	0	−1
Scissors	−1	1	0

The row that starts with the word Paper represents the results when Red chooses paper; similarly, the column headed Rock represents the results when Blue chooses rock. The number that is simultaneously in the Paper row and the Rock column is 1, which represents a gain to Red of 1 point when Red chooses paper and Blue chooses rock.

You can see that if the number 2 were in the Paper row and the Rock column, but all of the other numbers remained the same, it would make Red more likely to choose paper, because if Blue were to choose rock, Red would win 2 points. This change would also make Blue less likely to choose rock as well.

Mathematicians have devised a complete theory for analyzing what are called $m \times n$ games, where Red has a choice of *m* strategies and Blue a choice of *n* strategies. The mathematical analysis of such games is beyond the scope of this book (although a nice and eminently readable treatment of it appears in J. D. Williams's classic book *The Compleat Strategyst*; despite its title, it was written in the 1950s), but arithmetic alone will suffice to analyze a very important class of games, the 2×2 games, where each player has a choice of precisely two strategies.[1]

Third and Six

Over the years, football has become America's favorite sport; the Super Bowl attracts more spectators annually than any other single event on television. I'll assume the reader is familiar with the basics of football, but even if you've never seen an instant of a football game, the analysis is still easy to understand simply by looking at the numbers. Imagine instead that Red's three strategies in rock, paper, scissors were denoted Red 1 (the first row), Red 2, and Red 3, and similarly for Blue's three strategies. We know what the payoffs are when each player chooses a particular strategy, and that's all we need to know to analyze the game.

Let's look at a well-known situation in football: third down and six. The offense's goal is to make a first down, and the defense's goal is to prevent the offense from doing so. The offense has two basic strategies: to run or to pass. The defense has two fundamental strategies: a run defense (geared primarily to stopping an offensive run) or a pass defense (aimed mainly at stopping an offensive pass). The numbers in the following payoff matrix represent the percentage of times that the offense is successful, based on the strategy choices of each team. A football coach wishing to perform an analysis of this type would use percentages that are computed empirically, by looking at the records of past games, but the numbers here are chosen because they seem plausible and make for easy computation.

	Defense Strategy	
	Run Defense	Pass Defense
Offense Strategy		
Run Play	10	30
Pass Play	70	40

It doesn't take a rocket scientist—or a highly salaried football coach—to work out what's going to happen in this instance. The best that can happen if the offense chooses to run is that it succeeds 30 percent of the time. The worst that can happen if the offense chooses to pass is that it succeeds 40 percent of the time. Because the worst passing result is better than the best running result, the offense will *always* choose to pass.

Just as the offense wants to maximize the number of times it makes a first down—in other words, it seeks a strategy that results in the largest long-term payoff—the defense wants to minimize the number of times the offense makes a first down and looks for a strategy that results in the smallest long-term payoff. It cannot make the same type of analysis as the offense. Its worst result from employing a run defense (the offense makes a first down 70 percent of the time) is worse than its best result from employing a pass defense (the offense makes a first down 30 percent of the time). Also, its worst result from using a pass defense (the offense makes a first down 40 percent of the time) is worse than its best result from using a run defense (the offense makes a first down 10 percent of the time). The defense, however, is perfectly capable of analyzing the game from the standpoint of the offense, and it realizes that the offense will *always* pass. Knowing that the offense will *always* pass, it can choose its best strategy simply by seeking to minimize the number in the Pass Play row, and so the defense *always* adopts a pass defense on third and six. Each side is said to have adopted a pure strategy—by doing the same thing every time, rather than "mixing it up" as one does when correctly playing rock, paper, scissors.

When the offense always chooses to pass and the defense always uses a pass defense, the offense succeeds 40 percent of the time; the number 40 is called the value of the game.

There is an interesting aspect to this situation that deserves mention. Once the correct strategy is chosen by each side, any deviation from the correct strategy is punished. If the offense chooses to run while the defense is defending against a pass, its success probability is reduced from 40 percent to 30 percent. If the defense chooses to defend against a run while the offense elects to pass, the offense's success probability increases from 40 percent to 70 percent. Neither side has an incentive to change strategies.

If we switch the rows and columns of the matrix on page 26 (and change the game to a more abstract contest between Red and Blue), it would look like this:

		Blue	
		Blue 1	Blue 2
Red			
	Red 1	10	70
	Red 2	30	40

If we were to analyze this game from the standpoint of Red, there is no obvious strategy: the worst result of playing Red 1, 10, is less than the best result of playing Red 2, 40. Similarly, the worst result of playing Red 2, 30, is less than the best result of playing Red 1, 70. From the standpoint of Blue, however, things are much clearer: the worst result of playing Blue 1, 30, is better than the best result of playing Blue 2, 40—remember, smaller numbers are good for Blue. So Blue *always* plays Blue 1, and knowing this, Red will *always* play Red 2. The value of this game is 30.

In each of the two games discussed previously, one side has a clear choice: its worst result from playing one strategy is better than its best result from playing the other. In the matrix on page 27, if the number 30 were changed to 40, it would still be correct for Blue to play Blue 1, because its worst result from Blue 1 is at least as good as its best result from playing Blue 2. In analyzing a 2×2 game, the first step is to see whether one side or the other has a strategy whose worst result is at least as good as its best result from the other strategy. If so, the analysis proceeds in a straightforward fashion, with one player always doing the obvious thing and the other player reacting because he knows what the other player is going to do.

There is an alternative way to see whether one side or the other has a pure strategy. Let's take another look at the first case we examined.

	Defense Strategy	
	Run Defense	Pass Defense
Offense Strategy		
Run Play	10	30
Pass Play	70	40

From the standpoint of the offense, it is easy to see that it is better to pass than to run, no matter which defensive alignment the offense encounters. If it encounters a run defense, a pass succeeds 70 percent of the time, as opposed to the 10 percent of the time that a run succeeds. Similarly, if the offense encounters a pass defense, a pass is more likely to be successful than a run is. So a pass is clearly preferable to a run in either case.

Let's change the numbers a little.

	Defense Strategy	
	Run Defense	Pass Defense
Offense Strategy		
Run Play	50	30
Pass Play	70	40

In this case, the worst that can happen when the offense passes is not as good as the best that can happen when the offense runs, so on the basis of that criterion we cannot immediately say that the offense will always pass. When we examine things on a case-by-case basis, however, we see that a pass is always more successful than a run, no matter what defense is used, so the offense will clearly pass.

In the preceding diagram, the offense has a pure strategy because passing does better than running against each of the defensive options, although the worst result from passing is not better than the best result from running. If you look at the diagram at the bottom of page 28 from the standpoint of the defense, however, the worst result from employing a pass defense (the offense succeeds 40 percent of the time) is better than the best result from employing a run defense (the offense succeeds 50 percent of the time), so the defense will always employ a pass defense based on the criterion that its worst result from doing so is at least as good as its best result from using a run defense. It really doesn't matter whether you use the first criterion or the second to see whether there is a pure strategy—as long as you apply the criterion to both sides.

First and Ten

Another standard situation that recurs in football is first down and ten. Once again, the offense has the choice of a running play or a passing play, and the defense has the choice of which defense to use. The payoffs here are different, however; the offense seeks to maximize the average number of yards gained, and the defense to minimize this number. The payoff matrix for this situation looks like the following:

	Defense Strategy	
	Run Defense	Pass Defense
Offense Strategy		
Run Play	3	5
Pass Play	8	4

For the offense, the worst result of a run is 3, which is worse than 8, the best result of a pass. Also, the worst result of a pass is 4, which is worse than 5, the best result of a run. Looking at it from the standpoint of the defense, the worst result of a run defense is 8, which is worse than 4, the best result of a pass defense. Finally, the worst result of a pass defense is 5, which is worse than 3, the best result of a run defense. Alternatively, a case-by-case analysis shows no clear winner. Neither side has a pure strategy that it can adopt according to the guidelines we previously examined.

There is also a dynamic aspect to this game that differs from the situation we examined in third and six. No matter which strategies are selected by both sides, one side can always improve its position by changing strategies if the other one stays with its current strategy. For instance, if the offense chooses to run and the defense defends against a run (average yards gained = 3), the offense can improve its situation by deciding to pass while the defense still defends against a run (average yards gained = 8). The offense can improve its position when the payoffs are 3 and 4, whereas the defense can improve its position when the payoffs are 5 and 8. The same thing happens in rock, paper, scissors: no matter which strategies are selected by the players, one side can always benefit if the other continues to do the same thing.

The similarities continue between this game and rock, paper, scissors. In order to adopt the best strategy, each side must assume that the other side is a computer with perfect knowledge and must adopt a strategy that neutralizes the other's strategy. This can be done by making the long-term average payoff the same against either of the opponents' strategies; it's another place in which expected value appears.

Although a full analysis of this requires some algebra, this problem could have been handled in the eighth grade in 1895 Kansas simply with arithmetic.[2] Let's look at it from the standpoint of the offense, with the intention of first finding what percentage of the time it should pass and what percentage of the time it should run.

If the offense always chose to run and the defense elected to employ a pass defense, the defense could improve its result by 2 points by switching to a run defense. If the offense always chose to pass and the defense elected to employ a run defense, the defense could improve its result by 4 points by switching to a pass defense. Therefore, the offense should run twice as often as it should pass (the ratio of 4 to 2). To see that this "nullifies" any defensive strategy, let's assume the offense runs twice and passes once. If the defense chooses a run defense, the offense gains 3 yards twice and 8 yards once, a total of 14 yards in 3 plays, for an average of $4\frac{2}{3}$ yards per play. If the defense chooses a pass defense, the offense gains 5 yards twice and 4 yards once: again, an average of $4\frac{2}{3}$ yards per play. Therefore, $4\frac{2}{3}$ yards per play is the value of the game, the average that the offense will gain no matter what the defense does. Of course, as with rock, paper, scissors, the offense wants to choose what to do so as to prevent the perfect computer that is the defense from being able to obtain an advantage, so it must make sure that it chooses to run twice as often as to pass in a random manner. One way to do this is to look at the second hand of your watch just before deciding on the play: if it is between 0 and 39, run; otherwise, pass. Or look at the game clock. This is good training for coaches in clock management, which some coaches definitely need.

How to Play 2 × 2 Games

Let's boil it down to two steps.

Step 1. Check to see whether either side has a pure strategy. This can be done by comparing rows to see if the numbers in one row are larger than the corresponding numbers in the other. If this is the case, the row with the larger numbers is the pure strategy that Red (the row player) will use. If it is not, compare columns to see whether the numbers in one column are smaller than the corresponding numbers in the other. If they are, the column with the smaller numbers is the pure strategy that Blue (the column player) will use.

If there is no pure strategy, we move on to step 2. Let's go back and look at the examples in "First and Ten."

	Defense Strategy	
	Run Defense	Pass Defense
Offense Strategy		
Run Play	3	5
Pass Play	8	4

Step 2. Simply subtract the smaller number from the larger in each row, and put the result in the *other* row. I'll label this result the ORD for "other row difference." Here's the updated table:

	Defense Strategy		
	Run Defense	Pass Defense	ORD
Offense Strategy			
Run Play	3	5	4
Pass Play	8	4	2

The ORD tells you that the offense should run 4 times and pass 2 times—randomly, of course. This is the same ratio as running twice and passing once. If the defense employs a run defense, two runs and a pass will average $2 \times 3 + 8 = 14$ yards, an average of $14/3 = 4\frac{2}{3}$ yards per play. If the defense chooses to defend against a pass, two runs and a pass will average $2 \times 5 + 4 = 14$ yards, again with an average of $4\frac{2}{3}$ yards per play. Always remember to first check for the existence of a pure strategy!

It turns out that 2×2 game theory has a surprising number of applications in the world we live in. I'll show you a few of these in the remainder of this chapter, and other examples will pop up throughout the book.

Arrival Time

Here's a situation from a totally different area. You've got tickets to an upcoming event and have to pick up a young woman whom you hope will become your significant other. Should you show up on time or be fashionably late?

Obviously, it would be great if you and your hoped-for significant other were on the same wavelength. The best result would be if you both were on time, for in this case you'd be sure that you would arrive in time for the start of the performance. It's not quite so good if you are both fashionably late, but at least you can both laugh about it, as neither can blame the other. If you're on time and she's late, there's a chance you might not make the performance, but at least it's not your fault. The scenario most likely to lead to a disaster is if you're late and she's on time; I don't need to spell this one out.

I encountered this type of situation before I learned game theory, and because I come from the school of thought that puts punctuality on the same pedestal as cleanliness in regard to its propinquity to godliness, I never had any trouble making this decision. I was so well-known for always showing up five minutes early that whenever friends invited me to a party, they got in the habit of telling me to come half an hour later than everyone else. I also, it must be admitted, irritated a few potential significant others with this behavior; even my wife is somewhat less than thrilled by my obsessive punctuality. I could perhaps have done better (in scoring points with potential significant others, *not* in the selection of a mate) had I known how game theory tackles this particular problem.

Given the analysis I explained a while back, and scoring 10 for the best possible result and 0 for the worst, I might have constructed the following table:

	Potential Significant Other	
	On Time	Fashionably Late
Me		
On time	10	2
Fashionably late	0	6

Incidentally, there is an important difference between this game and football (from a game-theory standpoint), in that a potential significant other is not an opponent who is trying to defeat you. Game theory doesn't apply only to situations in which you have an opponent actively opposed to your interests; it is also useful in situations such as this, where you are trying to determine the best course of action.

It's easy to see that no pure strategy should be selected by either party, so it's time to move into the arithmetic. Here's the updated table:

	Potential Significant Other		
	On Time	Fashionably Late	ORD
Me			
On time	10	2	6
Fashionably late	0	6	8

This tells me to be on time 6 times and fashionably late 8 times, a ratio of 3 to 4. Remember—compute the ORD *only* if there is no pure strategy.

Let's see if that works. We'll assume you are on time 3 times out of 7 and fashionably late the other 4 times. The total number of points you will accumulate if your potential significant other is always on time is $(3 \times 10) + (4 \times 0) = 30$, an average of $4\frac{2}{7}$. The total number of points you will accumulate if your potential significant other is always fashionably late is $(3 \times 342) + (4 \times 6) = 30$, also an average of $4\frac{2}{7}$. So this strategy effectively neutralizes the random aspect of your potential significant other's behavior; in the long run, it doesn't matter what she does.

Unfortunately, I learned about game theory *after* I had adopted the strategy of always being on time. I don't know how things would have turned out otherwise, but I do know that I would have annoyed fewer of my dates by making them feel that they just had to be ready because I *always* showed up on time.

Valuable Cargo

An interesting application of game theory arose during World War II. It was often necessary to transport an object of considerable value (secret equipment, a high-ranking dignitary) from one place to another, and the cargo was considered so valuable that two planes were sent. The valuable cargo would be placed in the lead plane, so that the rear plane could give covering fire if the lead plane was attacked (guns mounted on fighters fired forward). After a while, the enemy picked up on the fact that a two-plane formation generally denoted valuable cargo in the lead plane, and it would concentrate its attack on that plane—which led planners to place the valuable cargo in the rear plane to divert the enemy. After a few lead planes were shot down and found to contain nothing of value, the enemy picked up on this tactic and started going after the rear plane, thus achieving a higher rate of success because the rear plane was more vulnerable. As might be expected, the valuable cargo then got placed in the lead plane.

Suppose we assume that the cargo always gets through if the enemy attacks the wrong plane; it has an 85 percent chance of getting through if it's in the lead plane and the enemy attacks that plane; but it has only a 65 percent chance of getting through if it's in the rear plane and the enemy attacks that plane. The matrix is therefore

Enemy Attacks

Cargo In	Lead Plane	Rear Plane	ORD
Lead plane	85	100	35
Rear plane	100	65	15

I've filled in the ORD column because the preceding description makes it apparent that there is no pure strategy for either side. The cargo should be placed in the lead plane 35 times out of 50, or 7 out of 10, and it will get through $((7 \times 85) + (3 \times 100))/10 = 89.5$ percent of the time.

As you can see, 2×2 games can be used in a lot of different situations.[3] They will continue to put in an appearance throughout the book.

3

How Math Can Help
Your Love Life

How do you know when he or she is "the one"?

• • •

Whom should you ask to the senior prom?

• • •

Why are women reputed to be fickle while men
are steadfast?

This chapter is, to some extent, written with tongue in cheek—but not entirely. The original title for this book, suggested by my editor, was *How Math Can Get You Laid*. It reminds me of the time I had to prepare notes for a class in functional analysis at UCLA and I titled them "Functional Analysis, Sex, and Violence—Part 1: Functional Analysis."

There's nothing wrong with a good come-on—but you will notice that the title of this book *has* been changed.

Even though some of the ideas presented in this chapter *might* be useful in improving your love life, I have to admit in all honesty that I have found that proficiency in mathematics tends to turn members of the opposite sex off, rather than on. This was so evident to me that in the 1960s I took to telling prospective dates that I was an itinerant poet (no lie, some of my better efforts appear on my Web site), rather than inform them that I was a professor of mathematics. There was one notable exception, however. After I became the graduate adviser in the Mathematics Department, the first person to walk through my door was Linda, a recent UCLA graduate with an excellent record and letters of recommendation from several faculty members whom I knew quite well. Linda and I were married four years later, so math did manage to fulfill the function suggested in my editor's proposed title—but I strongly doubt that many readers of this book will become graduate advisers in a mathematics department.

How Do I Know If He (or She) Is "the One"?

Every so often, lightning strikes, and two people know almost immediately that they are right for each other. My guess is that this is a fairly rare occurrence. You meet, you date, you spend time together, and love gradually blossoms. You remember your first love and the loves in between then and now, and you ask yourself the question that is the subject of this section.

The reason you ask yourself this question is that there's always the nagging thought that if you throw this particular fish back into the pond, you might be able to hook a more attractive specimen: a mate more compatible, better looking, richer—whatever you desire. The problem therefore arises as to whether there is a way to play the dating game so as to give yourself the best chance of finding your optimum mate.

I was first made aware of this strategy by Charles Brenner, who brought it up in conjunction with a fairly well-known math problem, the solution of which lies somewhat beyond the scope of this book.[1] Suppose a bunch of cards is lying on a table; each card has a number written on it, with the number side lying facedown so that you can't see it. You know how many cards there are, but you have no idea what numbers are written on them. They may be large or small, positive or negative. You are allowed to look at as many cards as you like, one at a time. Your goal is to choose the card with the highest number on it. The difficulty is that you can choose only the card that you think has the highest number on it while you are looking at it; if you put it aside to look at another card, you can no longer choose the card you set aside. One possible strategy is to look at a percentage of the available cards as a sort of database for comparison, and as soon as you find a number larger than the highest one in the database, you choose that card. The mathematical question is: what size should the database be to give you the best chance of choosing the card with the highest number? Obviously, if you use only 1 percent of the cards, you have a database that is clearly inadequate, and if you use 99 percent of the cards, there's a 99 percent chance that the card with the highest number on it belongs to the database, so you are no longer allowed to choose it.

How does this relate to finding "the One"? A potential candidate is like one of these cards; those who satisfy your criteria are analogous to cards with very high numbers, candidates who don't are obviously the low-numbered cards. You don't know how big the numbers can be—for a guy, you could meet a woman with the smarts of Marie Curie, the looks of Angelina Jolie, and the bankroll of Oprah Winfrey. What you do know is approximately how long your prime dating lifetime is: if you are now twenty, you could be dating up until forty-five or thereabouts, but after that your chances are probably going to diminish materially. The length of your prime dating lifetime is the analogue of the number of cards on the table, and the comparison database is the length of time you will allow yourself to check out possibilities before saying, "As soon as I meet someone more appealing than anyone in my comparison database, I'll do my best to make this person my spouse."

Of course, the analogy between the card problem and the dating problem is not an exact one, but mathematics doesn't have to be perfect to have value—it merely has to be useful. For those waiting breathlessly to find out the length of time you should use to assemble your comparison database, it turns out to be the fraction $1/e$ ($e = 2.71828 \ldots$, the base of the natural logarithms) times the length of your prime dating lifetime. Thus, $1/e$ is approximately 37 percent; if you are twenty years old and figure that your prime dating lifetime is until you're forty-five, then 37 percent of twenty-five years is a little more than nine years. So you date until you're twenty-nine, and as soon as you find someone who's more appealing than anyone you dated between twenty and twenty-nine—go for it.

I'm not sure that Charles took his own advice; he married for the first time (and, so far, the only time) when he was in his early fifties. Of course, that might have been in accordance with this particular strategy if he considered his prime dating lifetime to be from ages twenty to ninety.

It's important to realize that this strategy is designed to maximize your chances of marrying the best available candidate whom you may possibly meet. If you simply wanted to maximize your chances of getting married, there are all sorts of ways to do this, especially since marriage to a U.S. citizen is highly coveted in many parts of the world.

This strategy also applies to other situations where you can assemble a database but must make a go or no-go decision each time a new item is added to the database. One such example is the purchase of a house. Just as in romance, you may walk into a house and know instantly that it is the one for you, but the chances are that you will probably have to accept one that is less than ideal. A good way to do this is to allow yourself a certain amount of time to become familiar with what's on the market, then pounce when you see a house that you feel is superior to those you have already examined.

One of Those Days

If you're a guy, something like the following scenario is almost certain to happen to you. You're walking home, and all of a sudden you realize that this is one of those special days: your wife's birthday, your wedding anniversary, the anniversary of the first time the two of you kissed. Whatever it is, your wife considers its celebration of major importance. Unfortunately, you simply can't remember whether today is one of those days—or not.

You pass a florist, and a bouquet of red roses beckons invitingly. Yet red roses have a different message before you are married and after. Before you are married, red roses say "I love you." After you are married, red roses say either "I love you," or "I screwed up, please forgive me." If you bring home a bouquet of red roses and it's one of those special days, you will probably be greeted by the phrase "Darling, you remembered!" or some variant, and you can buy time in order to discover exactly which one of those special days it is. On the other hand, if it's *not* one of those special days, there's a reasonable chance that the remainder of the evening will be devoted to your wife's trying to dig up exactly what you screwed up. You can rest assured that your explanation of "I couldn't remember whether today was [fill in appropriate special day here]" will almost certainly not be believed.

Your first thought might be, There are 365 days in the year and there are only a few special days, so this is very unlikely to be one of them. Yet the fact that you thought that today *might* be one of those special days considerably raises the probability that it *is* one. With no PDA or other resource to fall back on, you rely on game theory.

Obviously, you'll score maximum points if today is a special day and you bring home the bacon—I mean roses. It's a disaster if it's a special day and you forget, a lesser disaster if it's not a special day and you show up with roses, and it's just another day if indeed it is just another day, minus roses. That leads to the following diagram:

	Special Day?		
	Yes	No	ORD
Bring Roses?			
Yes	10	−4	10
No	−10	0	14

Of course, you always check to see whether there's a pure strategy, but there isn't, so I've included the ORD. The odds are 7 to 5 against your bringing roses, and the game has a small negative expectation of $-1\frac{2}{3}$. Well, what did you expect? If you were even in doubt, you were starting from behind the eight ball.

Prom Date

If you're going steady (or whatever the current expression is), the senior prom presents no obstacle other than what you should wear (if you're a girl) and what type of transportation you should provide (if you're a guy). If you're not going steady, however, there are all sorts of problems related to whom you should ask (if you're a guy) and what you should do if the wrong guy asks you (if you're a girl). Fortunately, mathematics can help you in either case.

Let's first look at the situation if you're a guy. You have a choice of two girls to ask to the prom, whom I will label "A-list" and "Backup." A-list is in high demand, and Backup is a good alternative; however, you have to decide which one to ask first. Obviously, you would prefer to ask A-list and have her accept, but if she turns you down not only are you really depressed, but someone else may have already grabbed Backup—and you're out in the cold. If you ask Backup first, there's a reasonable chance that she'll accept, but if she doesn't, it's not so bad because there's still a chance with A-list, who is known to be very choosy. After weighing all of this, you construct the following matrix.

	First Girl Asked Accepts	
	Yes	No
First Girl Asked		
A-list	10	0
Backup	6	4

Surprisingly enough, when you consider this from the standpoint of game theory, there is a pure strategy for the first girl whom you ask: she will probably turn you down. Consequently, you should ask Backup first. Here's the interesting part, though; Backup is usually aware that she's the second choice, and she will often accept, feeling that this is her best shot. This may explain one of life's more counterintuitive observations: the class hotties don't always show up at the senior prom!

Let's take a look at the other side of the equation (everyone *except* math teachers talks about the other side of the equation when there really isn't an equation; this gives me an opportunity to indulge in something that is usually anathema for math teachers). A-list has to decide whether to accept a prom invitation if she gets one from a boy who is reasonably attractive. After all, there's always the possibility that the grass is greener on the other side of the fence, and she may receive a better offer. As against that, if she turns down the invitation, she may find herself out in the cold, and tongues will *really* start to wag.

	Is There a Better Offer?		
	Yes	No	ORD
Accept Invitation			
Yes	5	8	10
No	10	0	3

Here, finally, may be the mathematical explanation of why men are thought of as steadfast and predictable, while women are fickle. As we have seen, guys have a pure strategy in issuing invitations to the senior prom, whereas girls should play a mixed strategy. Although the numbers may change, depending on the relative values that the girl attaches to having a date and staying home, all that will do is alter the relative frequency with which the invitation is accepted.

The Expected Value of Internet Dating

I remember seeing an episode of James Burke's wonderful television series *Connections*, in which he brought up the fact that one of the great benefits of rail transportation was the increase in genetic diversity of the human species.[2] True, soldiers and adventurers since the beginning of time had married women who were geographically inaccessible to most of their brethren, but the train made it possible to expand one's circle of potential mates to far beyond the confines of one's home village.

As time passed, the confluence of the ease of transportation and Internet dating not only improved the genetic diversity of the human species, it made it much easier for an individual to find a wide choice of potential mates. While admitting there are potential hazards to Internet dating, if I were single I'd be all over this innovation. I actually tried computer dating in its primitive form in the 1960s, and it didn't work out all that badly. From the standpoint of expected value, Internet dating is a huge winner. In some instances, there's no expense at all in the initial phase—several firms currently allow you to take a personality test and let you preview your potential matches. I cannot imagine why anyone who is single would not be willing to do this. Whatever your probability of success is, your reward for success is large, and your risk for failure is very small. Think about "It is better to have loved and lost than never to have loved at all," and you'll be taking those personality tests as soon as you put down this book.

A Question of Percentages

For my first date with Linda, we went to a Thai restaurant. As we happily munched away on the mee grob, I asked her what she thought made for a successful relationship. Without hesitation, she answered that the couple had to have 70 percent of things in common and 30 percent different.

I liked everything about this answer. First of all, percentages (as you know by now) are a big deal with me, so the mere fact that she phrased the answer in terms of percentages appealed to me. Second, she got the numbers right—I tend to like the familiar, but I'm at least somewhat open to new experiences.

Originally, I thought that the 70–30 ratio was right for everybody, but I've rethought my position on this subject. The ratio is a measure of how comfortable you are with the familiar, as opposed to how much you appreciate new and different experiences. To translate that number into the arena of cuisine, the 70–30 ratio that so appealed to me would clearly be much too adventurous for someone for whom pizza is exotic, and it would be much too conservative for someone willing to go to an Asian fusion restaurant and sample the insect dishes on the menu.

I think that you would benefit by figuring out what ratio works best for you, because I can't imagine that two people who have radically different views in this area will be happy with each other. I looked at Eharmony.com, a leading computer matchmaking organization, which measures twenty-nine dimensions of compatibility. None of those twenty-nine dimensions seemed to measure the ratio of what you have in common to what you don't. I certainly wouldn't put the ratio of this dimension near the top in choosing a mate (all of the obvious types of compatibility are much more important), but I think it represents a source of potential friction down the line. It pops up in numerous areas—what foods you eat, what entertainment you pursue, where you want to go on vacation. These things may not matter during the initial rush of excitement in getting to know someone, but they will later down the line.

4

How Math Can Help You Beat the Bookies

Why should your lottery ticket contain numbers greater than 31?

• • •

Can you overcome a negative expectation?

• • •

When should you bluff and when should you fold?

M any people find gambling extremely exciting, and as long as it's not done to excess, it's competitive in price with many other forms of entertainment. How much does it cost for one person to go to a first-run movie and have a tub of popcorn? As of this writing, in Los Angeles the cost is about $14 (when I was growing up in a suburb of Chicago, you could see a Saturday matinee double-feature with a cartoon and a newsreel, plus get popcorn, all for 35 cents—and the movies were a lot better than they are today). As you will see shortly, if you bet the line in football, your expectation is −4.54 percent. Therefore, if you bet games at random and wager a total of $300, it figures to cost you $13.62. The cost is basically the same, and for my money, I find gambling a lot more enjoyable than the latest overhyped, computer effects–studded, dialogue- and character-deficient film. Of course, that's a matter of taste, but if I bet $20 apiece in fifteen football games, I figure to get more than forty hours of nail-biting excitement from rooting for my money.

This chapter may not turn you into a winner, but it will certainly help you lose less in the long run. I'd rather see you get the maximum entertainment value for your gambling dollar, and the best way to do this is to show you how to gamble in an intelligent fashion.

Regard gambling as entertainment, and set a reasonable limit for the amount you are willing to lose each month. When you hit that limit, *stop betting for the month*. Or do it on a weekly basis, if you can't stand the thought of hitting your limit by the tenth of the month and going without any "action" for three weeks.

The Three Types of Games

As I see it, there are three types of games at which you can gamble:

- Games where the odds are stacked against you
- Games of skill
- Games in which you can overcome an ostensibly negative expectation

Games Where the Odds Are Stacked against You

The first type of game is one you cannot win in the long run. Games in this category include most of the standard casino games such as craps, roulette, slot machines, and keno. These games have a negative expectation that no strategy, short of cheating, can overcome. We saw an example of this in chapter 1, when we looked at the expectation of a bet on red in American roulette. Similar situations exist in all casino games: the payoff and the probabilities are adjusted so that all bets have a negative expectation. There used to be one exception: blackjack. By keeping track of the cards, it is possible to determine when the player's expectation is positive, and one can bet more in such situations. Casinos know this, however, and either use automatic shufflers, which negate this advantage, or politely ask you to leave if they can tell by your play pattern that you are successfully counting cards.

State lotteries are slightly different, because it is possible to play state lotteries intelligently. Of course, we need to calculate the expectation of a lottery ticket. If you play a particular lottery, you can usually find the number of different possible tickets you could buy—these are computed by techniques referred to as combinatorics and are often found as problems in second-year high-school algebra.

The number of different possible lottery tickets is computed using what is known as the combination formula: a lottery ticket in which you choose six numbers from the numbers 1 through 51 is called a combination of 51 things taken six at a time. The number of such lottery tickets is written $C(51,6)$ and is computed by means of the formula

$$C(51,6) = (51 \times 50 \times 29 \times 48 \times 47 \times 46)/$$
$$(6 \times 5 \times 4 \times 3 \times 2 \times 1) = 18{,}009{,}640$$

For the curious, a (very) short course in basic combinatorics can be found at the California State University at Long Beach Web site, which I used for an explanation of expected value.[1]

If the jackpot is $10,000,000, most states pay that off over a number of years. To compensate for the fact that inflation takes a bite out of payments made in the future, multiply the number of possible tickets by 2.5 (the math behind this will be explained in detail in chapter 8). Then take your tax bracket into account—your winnings will lift you into the highest tax bracket, which we will assume is 50 percent. To compensate for the share of your winnings that will be heartlessly ripped away by the tax agencies, multiply by 2 the number you obtained as the result of multiplying by 2.5. If the jackpot is higher than that amount, buy a ticket. You generally have a positive expectation (but don't expect that positive expectation to correspond to money in the bank—a positive expectation on an event that happens one time in fifty million is quite different from a positive expectation on an event that happens one time in three).

In addition, it would be terrible if you finally hit the jackpot, only to discover that your prize had to be shared among other winners. Well, it wouldn't *really* be terrible—you still won the lottery—but it would be nice if the only person you had to share your good fortune with was Uncle Sam. In order to minimize the possibility of sharing with other winners, be aware that many people embed lucky numbers in their tickets: the day they were married, the day their first child was born, and so on. To avoid having to share your largesse with these people, use numbers larger than 31 to construct your lottery ticket.

Games of Skill

A game of skill is one in which a good player has a positive expectation. It is possible to play winning blackjack in a casino (although, as previously mentioned, you must be a *very* good player, and you must play in casinos that don't use automatic shufflers), and it is possible to win certain other games, such as poker or backgammon, through expertise. Many people gamble at golf, which also falls under this category. When I was younger, I was a reasonably successful gambler, but I stuck to poker, backgammon, and counting cards at blackjack, because I felt that my ability to calculate probabilities and expectations would enable me to win. For the most part, it did—it wasn't that I was so incredibly good at this, but that most people (or, at least, the people I played with) were considerably worse.

I also found that it was much easier to win at backgammon than at poker, for several reasons. One is that backgammon is a game with very little in the way of psychology, whereas everyone knows that much of poker is psychology: bluffing, and reading your opponent. More important, though, is that the difference between a winning player and a losing player is that the winning player makes better decisions than a losing player does.

In an evening of poker, you may have five or six crucial decisions that will determine whether you are a winner or a loser. Generally, those key decisions come when you have a good hand, someone else also has a good hand, and a disproportionate amount of money is bet relative to the amount that would be bet on a routine hand. Books have been written on this subject, but the most valuable secrets are generally learned only through experience—and those who have learned them best are often the winners of the multimillion-dollar poker tournaments.

If you play backgammon for the same amount of time, however, you may have to make hundreds of decisions; the evening's results depend more on making a lot of little decisions consistently better than your opponent does. A good backgammon player will win much more frequently against weak competition than a good poker player will win against the same, and psychology (one of my areas of weakness) doesn't enter into backgammon anywhere near the extent that it does in poker. I played backgammon to support myself during the year that I could not find a job teaching mathematics, and my opponents often made such poor moves that it was like dropping rocks and watching the other players bet that the rocks would fall up rather than down.

Games in Which You Can Overcome an Ostensibly Negative Expectation

There are a few games in which you theoretically have a negative expectation, but that expectation can be overcome. Much of sports gambling is of this type. The vast majority of sports gamblers lose, but there is a small minority who win, and win consistently.

I have known a few very successful sports gamblers (and many unsuccessful ones). Most of the successful ones had some sort of underlying strategy that they used, and they coupled this with self-discipline. Just because the Super Bowl is the biggest game of the year, if you don't see a clear-cut favorite don't feel that you simply have to bet because it's the biggest game of the year.

I cannot promise that I can make you a winner. But I can promise that I can make you less of a loser, and I will show you what *can* make you a winner. That's basically what this chapter is about.

Calculating the Bookie's Edge

The neighborhood bookie of my youth has largely been replaced by an impersonal casino located somewhere in cyberspace, but most of the betting is still the same. In order to start losing less, you must know what you are trying to overcome, and that means being able to calculate the house percentage of a bet. The house percentage of a bet is similar to percentage expectation but has a slightly different meaning, because it is not possible to know the probability that a given event will occur. One can calculate the exact probability of getting dealt a blackjack or rolling a 7 at the craps table, but one cannot calculate the probability that the Dallas Cowboys will beat the New York Giants on Sunday.

Let's start with the simplest bet, the 11–10 pick 'em, which is the backbone of the industry. This is the standard football line bet. If you see a game listed as Dallas −3 over the Giants, this means that you must wager $11 to win $10, and in order for a bet on Dallas to win, Dallas's score must exceed that of the Giants by more than 3 points. If you bet on the Giants, you win as long as the result of adding 3 points to the number of points scored by the Giants exceeds the number of points scored by Dallas—but you almost certainly knew that. If Dallas beats the Giants by exactly 3 points, the bet is a push: no money changes hands.

The 11–10 number, with the loser paying the 11, has been the industry standard for many years. Obviously, the procedure must be designed so that the entity offering the opportunity to bet has a positive expectation—as with insurance, that entity is taking a very large risk, whereas you are risking only the amount you lose. That entity is also offering you a service, and one must pay for a service. The 11–10 ratio is brilliantly chosen, because it eats up the loser's money slowly, offering the illusion that if only that last-minute field goal hadn't gone wide, he'd be a winner on the day. This ensures that the loser always returns, generally to lose again.

To calculate the house percentage of this bet, let's see how much you have to wager in order to receive $100 back from the bookie. A wager that wins will invest $11 to win $10, so an investment of $11 will result in the bookie paying you back $21, the $11 you bet plus the $10 you won. If you wish to receive $100 back from the bookie after a winning bet, realizing that $100 = 4.762 × $21, you would have to bet 4.762 × $11 = $52.38. If you were to bet $52.38 on both the Giants and Dallas, you would lose one bet and receive $100 back from the bookie. Because 2 × $52.38 = $104.76, you must bet $104.76 to receive $100 back. You would therefore lose $4.76 on a wager of $104.76, or 4.54% of the amount that is bet. This is the house percentage. If an equal amount of money is wagered on the Giants and the Cowboys, the house would take 4.54% of the amount that is bet.

In fact, that's the goal of setting the line: it does not represent the estimate of what will happen, but it represents the house's estimate of what number will attract an equal amount of wagering on both sides. That way, the house has a sure win. If most of the money is wagered on Dallas, the house has a risk if Dallas turns out to be a winning bet. Even though, in theory, the house percentage should give the house a win in the long run, the house would much prefer a sure win in the short run. Who wouldn't?

This same type of line exists for basketball games ("Spurs +5 over the Bulls"), and over-unders in all sports. The expectation is the same −4.54%, as long as you must pony up $11 to win $10. Just as you can shop for bargains at the supermarket, however, you can shop for bargains at the cybernet casino. You will often find different casinos giving specials of 21 to 20, and if you perform the same calculation, you would find that you need to bet a total of $102.44 to win $100. This means you would lose $2.44 on a wager of $102.44, or 2.38% of the amount that is bet. This is a lot better than a typical wager you can make in the California state lottery and much better than almost anything you will find in a casino in Las Vegas or Atlantic City.

A relatively recent development has been the advent of betting exchanges. Betfair is the industry leader. Nominally, Betfair cannot be accessed from an IP address in the United States, but wherever a demand exists, there's a way to circumvent this—for a price, of course. A betting exchange allows the public to set the odds. If the casinos have the line on the Patriots-Cowboys game as Patriots −3, and you are willing to take the Cowboys if you receive 4 points rather than 3, you can place an order on the betting exchange to do so, much as you can place an order to buy a stock at a price currently below the market. Just as your stock order may go unfilled, so may your order on a betting exchange. The betting exchange will take a percentage of your winning bet (good customers get better rates), but in general you get a better deal at a betting exchange than at an Internet casino that offers odds.

In addition to the standard line bet, there are three other types of commonly available bets. I'll discuss each one separately, because the computation of the house percentage varies for each, although the idea is similar.

Bets Made with Odds

The most common example of bets made with odds is horse racing, where you bet horse number 5 at odds of 7 to 2: you win $7 for each $2 bet if horse number 5 wins the race. Another variation of this can be found in football or baseball. At the start of the season, odds will be posted for all of the teams for winning the Super Bowl. You can find variations of this bet; for instance, you might find a bookmaker who offers odds that the San Diego Chargers will win the AFC West.

To illustrate the idea, let me make up an imaginary odds chart for the AFC West.

Team	Odds
Kansas City	4–1
Denver	1–1
Oakland	3–1
San Diego	3–2
Seattle	9–1

Let's see how much we would have to bet on each team to receive $100. A $20 bet on Kansas City at odds of 4 to 1 would win $80; we'd get $100 back (the $20 we bet plus the $80 we win) if Kansas City wins the AFC West. Similarly, we'd need to bet $50 on Denver to get back $100, $25 on Oakland, $40 on San Diego, and $10 on Seattle. That's a total of $20 + $50 + $25 + $40 + $10 = $145.

If we made all of those bets simultaneously, we'd bet a total of $145, and no matter who won, we'd get back $100. If we bet $100, we'd win 100/145 × $100, or $68.97. So for every

$100 bet, we would lose $100 − $68.97 = $31.03; therefore, the house percentage is −31.03%. Brrr!

You've got to be an absolute masochist to bet money with an expectation of −31%, especially when there are plenty of good bets available at −4.54%. The sad news is that horse racing generally has house percentages in the −20% range, and betting at the start of the season on who will win the Super Bowl usually has an even worse house percentage.

There's a formula for computing the expectation given all the odds, but why bother? (This is one of my favorite lines as a teacher: "There's a formula, but why bother?") If you understand the basic idea, you can always work it out.

I long for the day when I see odds posted like this:

Team	Odds
Denver	4–1
Kansas City	4–1
Oakland	3–1
San Diego	4–1
Seattle	9–1

In order to get back $100, bet $20 each on Denver, Kansas City, and San Diego; $25 on Oakland; and $10 on Seattle, for a total of $95. Make sure that you will get paid when you win, because a bookie who offers odds like this generally has a one-way ticket to Rio de Janeiro in his pocket or is operating from south of the border to begin with.

Baseball Odds

Most of the betting in baseball is done via an odds line. If the Giants are playing the Dodgers in L.A., the odds might be quoted as "Dodgers $1.30−$1.50." This means that if you want to bet the Dodgers (the favorites), you have to bet $1.50 to win $1.00; and if you want to bet the Giants, you bet $1.00 to win $1.30. To compute the expectation, we do the same thing as for horse racing or betting who wins the AFC West. In order to get back $100 by betting on the Dodgers, we must bet $60: we'll get back $100 by receiving the $60 we bet plus the $40 we win. Similarly, to get back $100 by betting on the Giants, we must bet $43.48. So if we bet $60 + $43.48 = $103.48, we'll get back $100, no matter who wins. Alternatively, we could bet $100 and get back $96.64, so the house percentage is 3.36%.

Need I say that this is greatly preferable to a percentage expectation of −31%?

Parlays

Many bookies will offer you the following arrangement, which is known as a parlay. Bet any two games against the line. If you win *both* games, you will win $13 for every $5 you bet. If you don't win both games, you'll lose $5.

We can't compute the house percentage on a parlay bet because the house won't let us take the other side of the wager, where we lose only if we lose both games. What we could do is make our selections in both games by flipping a fair coin. There are four possible ways the two flips can come up, and each is equally likely: heads-heads, heads-tails, tails-heads, and tails-tails. Only one of these combinations will win for us, so if we play this game four times, we will have wagered 4 × $5 = $20 and ended up losing $2 (we won $13 once and lost $5 three times). This is a percentage expectation of −10%; the percentage expectation for a straight bet is generally −4.54%. It's not as bad as horse racing, but it's still a sucker bet.

Teasers

There are all sorts of variations on teasers, but they work basically like this. The line on the first game of the University of Southern California's 2008 schedule is an August 30 matchup with Virginia, and USC is a 20-point favorite. Suppose that you are offered a 6-point teaser by laying odds of 2 to 1. The bookie will allow you to take 6 extra points; if you bet on USC, you must give 14 points (rather than having to give 20 points), and if you bet on Virginia, you will receive 26 points (rather than receiving 20 points). Instead of laying 11 to 10 on a losing bet, however, as you do when in a standard bet against the line, you have to lay 2 to 1.

In general, it is a reasonable assumption that a bet against the line has a probability of ½ of winning, but one cannot estimate the probability of winning a teaser bet without knowing the distribution of a specific random variable—how the real world does against the line. If the line was so accurate that 40 percent of the games ended up within 6 points of the line, you'd "tease" both sides of the bet, winning both bets if USC won by between 14 and 26 points, and winning only one of the bets the other time. If you did this for five games, betting one unit on each team, you'd win two units twice and lose one unit three times, for a pretty hefty return on capital.

To compute your expectation accurately, you'd have to know the distribution of how games ended up versus the line. I don't know this, but my smart gambling friends tell me that teasers are a sucker bet, and I trust them. Incidentally, the definition of a sucker bet is basically one with a large house percentage.

Can You Win at Betting Sports?

The answer to this is a heavily qualified "yes." The first thing to do is make sure to avoid sucker bets: bets with large house percentages. I think that a good rule of thumb is not to make any bet with a house percentage of more than 5%, and most smart gamblers would agree (even though there are some who can consistently beat the horses, despite a house percentage of close to 20%).

Recall that the line is basically an estimate of what the public thinks. If the line is accurate in attracting half the action on each side, when you bet against the line you are backing your judgment against that of the public, with the bookie merely an intermediary who facilitates this. Therefore, if you can estimate a team's chances of beating the line better than the public can, you can win at sports betting.

Middling the Line: One Possible Way to Win

Arbitrage is a fancy name for buying the same commodity at a lower price than you sell it for and is a time-honored way to make money. If gold is selling for $950 an ounce on the London exchange and you can buy it on the New York exchange for $940, an arbitrageur will buy a gold contract in New York for $940 an ounce and sell it in London for $950 an ounce, making the difference of $10 an ounce. This is harder to do in the Internet era than it was a generation or two ago, because price disparities are observed by a large number of people and therefore quickly disappear, but that's the idea.

The analogy in sports is to find different bookmakers who offer different lines on the same game. Suppose that BestBet lists the Patriots as a 3-point favorite over the Cowboys, whereas WagerWorld has the Patriots favored by 3½ points. Let's assume for the moment that each line is the standard 11–10. If you bet $100 on the Cowboys at WagerWorld and $100 on the Patriots at BestBet, if the game ends with any score other than the Patriots winning by 3, you will win one bet and lose the other. You'll make $100 on your winning bet and lose $110 on your losing bet, for a net loss of $10.

Let's suppose, however, that the final score is Patriots 17, Cowboys 14. You win $100 at WagerWorld but tie the game at BestBet, and ties do not result in any money changing hands. As a result, you win $100.

The question then becomes one of probability. If the game ends with the Patriots winning by 3 more than 1 time in 11, you will end up a winner; less, and you will end up a loser. There are ways that you can improve on this.

The two basic ways to do so are by getting better prices or more disparity in the two lines. If you are able to get odds of 21–20, rather than 11–10, you will only lose $5 on any game that doesn't, as the gamblers say, "end on the number." As a result, you need only win 1 game in 21 to break even. If you can find a bookie who has the line Patriots −2½ to go with the line at WagerWorld of Cowboys +3½, even if you assume that you are laying 11–10 odds on both bets, you will win $200 if the game ends with the Patriots winning by 3, as opposed to losing $10 for any other result. This particular example clarifies the term "middling the lines," because the game "ends on the number" in the middle of the range from 2½ to 3½.

Middling the lines in football is slightly different from middling the lines in baseball or basketball, because in football some numbers are more "live" than others. For example, 3 is a live number because a lot of close games are decided by precisely the value of a field goal—3 points. But 2 is nowhere near as live; simply check out the scores on any Saturday or Sunday to see how many more games end with a point difference of 3 than end with 2.

Middles exist because there is not a single monolithic entity setting the line, just as the price of a box of Kellogg's Corn Flakes differs slightly from supermarket to supermarket. The opening line—the first line offered by the bookmaker—may differ because each casino has a different way of assessing the number that will attract equal betting on both sides of the line, the "Holy Grail" of the bookmaker. In addition, the vagaries of the clientele wagering at each particular casino may generate an imbalance. Suppose that the opening line at BestBet has the Patriots favored by 3 points over the Cowboys—and a lot of money is bet on the Cowboys. In order to "balance" their action, BestBet needs to attract more Patriot bettors, and it can do so by lowering the line from 3 to 2½.

Just as there are commodity arbitrageurs who get "lines" from all of the major exchanges in the world and make a healthy living on the price differences, there are gamblers who have accounts at every casino and make a living by middling the lines. I didn't write this book to persuade people that this would be a good way to make a living, but I think I've performed a socially beneficial act if someone reads this book and is converted from being an "action junkie" who bets every game and every hunch to a "shopper" who hunts for betting bargains.

Advisory Services

Possibly the only people who regularly appear on TV and are sleazier than politicians are the ones who offer to sell you information on upcoming games. Despite the fact that I am allergic to people who have slicked hair, wear cheap sports coats, and look like Vegas pit bosses, I am willing to admit that some of them may actually be able to deliver the goods.

You're going to have to do the work, though. One thing to be wary of is the claim that these people are 12 and 2 in the last two weeks. How is this documented? Are they picking against the spread or the money line? The money line is an odds bet; in the USC-Virginia game described previously, the money line is a baseball-type line in which you can either give odds (by betting on USC) or receive odds (by betting on Virginia). This game might be listed as $11−$10; you need to bet $11 on USC to win $1 if USC wins the football game, and if you bet $1 on Virginia to win the game and it actually does (lots of luck on this one), you will win $10. Going 12 and 2 on money-line bets is very different from going 12 and 2 on bets against the line. It's up to you to find out; I have better things to do with my time.

Calling a Bluff Using Game Theory

Although a familiarity with game theory won't turn you into a winner, it's a useful tool to have at your disposal, especially when you're gambling in situations where bluffing plays a role, such as poker.

Nelson Algren, the author of *The Man with the Golden Arm*, once propounded three rules for living: never eat at a place named Mom's, never play poker with a man named Doc, and never sleep with someone whose troubles are worse than yours. Two out of three ain't bad, you muse, as you contemplate a reasonable hand in Texas Hold 'Em, but Doc looks you in the eye and shoves in a large bet. You figure out that if you call and he is bluffing, you'll win $700, but if you call and he has the goods, you'll lose $300. On the other hand, if you fold and he is bluffing, you'll lose $200, but if you fold and he has the goods, you'll break even. What to do?

You quickly excuse yourself, ostensibly to go to the men's room, where you take out a pencil and paper and scribble the following matrix:

	Doc Holds	
	The Goods	Big Bluff
Your Action		
Call	−300	700
Fold	0	−200

You quickly realize that neither you nor Doc has an obvious pure strategy, so you undertake the usual analysis.

	Doc Holds		
	The Goods	Big Bluff	ORD
Your Action			
Call	−300	700	200
Fold	0	−200	1,000

You have to call 200 times and fold 1,000 times, a ratio of 1 to 5. To see whether this is the correct ratio, imagine that Doc holds the goods, and you call once and fold 5 times. You'll lose $300, an average loss of $50 per game. If Doc is running a bluff, you'll win $700 once and lose $200 five times—again, an average loss of $50 per hand.

Not having a die handy, you use the second hand of your watch as a randomizing device, deciding that if the second hand is between 0 and 9, you'll call, and you will fold otherwise (there are 10 numbers between 0 and 9 and 50 between 10 and 59). You glance at your watch, go back to the room and toss in your hand. Hey, this is real life, and the important thing is to make the play that is a long-term winner. As Kenny Rogers said, you've got to know when to fold 'em, as well as when to hold 'em. Only in fiction does Doc's eyelid twitch when he is bluffing.

When to Bluff

Again, volumes have been written on this subject. There are basically two ideas behind bluffing: to win a particular bet (the bluff), and to make your opponent think you could be running a bluff so that he will call your large bet when you actually have the goods. A bluff of the latter type is an investment, not unlike advertising.

Let's see how game theory might handle a particular situation. You've invested $100 in a pot and your flush draw just busted, leaving you with jack high. If you bet $300 and your opponent folds, you'll win $200. If he calls, however, you're out a total of $400.

	Doc's Action	
	Call	Fold
Your Action		
Bluff	−400	200
Fold	−100	−100

As you can see, calling is at least as good as folding for Doc on a case-by-case basis, so he'll always call—and thus you should fold. There's an argument here that if you fold, he doesn't have the choice of calling or folding, but you could modify your choices to bluff and show or fold, and Doc's to call and show or fold.

Yet if you can estimate (based on previous experience or your innate ability to estimate these things) how likely Doc is to fold, you can determine by using expected value whether it's a good idea to bluff. It's much like insurance: if you don't bluff, you are guaranteed to lose $100, but if you feel your opponent will fold more than half the time, it's a good bet. After all, if you bluff twice and your opponent folds once, you make $200 when he folds but lose $400 when he calls, an expectation of −$100. Because you will usually lose more when an opponent calls a bluff than you will win if he folds (a bet of $300 isn't going to scare him out of a pot with $10,000 in it), bluffing isn't a winning tactic on any individual hand. But if you view it as money invested to get your opponent to call your good hands, it could be.

5

How Math Can Improve Your Grades

Will guessing on a multiple-choice test get you a
better score?

• • •

What test subject should you spend the most time
studying for?

• • •

What subject should you major in?

I've been teaching college for nearly forty years, and it amazes
me how little attention students seem to pay to obvious
methods of improving their grades. I'm not talking about
the straightforward approach of spending more time on studies
and less time on texting or Facebook. I'm talking about getting

better grades simply by employing a little ordinary intelligence. Most of these techniques are related to the mathematics involved in taking tests and allocating the time available for studying, but here's one piece of advice that is pretty generally applicable: unless you are studying something that you find absolutely fascinating (like *that* ever happens in school), *never* study for more than two consecutive hours. I tell my students that when they are doing math problems, they'll do a pretty good job the first hour, maybe an okay job the second hour, but by the third hour their brains will have turned to tapioca. Take a break. A 2004 study published in the journal *Pediatrics* cited excessive TV watching as a possible cause, but one thing is clear: if you're not paying close attention, you're not learning efficiently.[1]

Strategies for Taking Tests

I should say at first that most of the information in this section applies to tests that have some sort of numerical scoring: problem-solving tests such as you will find in math and science classes, multiple-choice tests, and true-false tests. You might be able to apply one or two tips to papers you write or essay tests that you take in class, but that's not the type of test for which math can be of substantial help.

It may seem too obvious to mention, but you have to know how a test is being scored. Math and science tests either have all problems given equal weight, or various problems have different point values, and the point values are clearly indicated on each problem. It also obviously helps to know your teacher's preferences; most teachers try to test the material they consider to be the most important. When I am teaching any course except upper-division math courses, I emphasize practical problems and story problems, and I tell this to students upfront on the first day of class and in the course syllabus. Some teachers are more subtle than I am about this, but you can read the lips of most of them. If a teacher says this is important, the odds are strong that he or she will test it. In courses such as math or science, if you know your teacher has a preference for certain types of problems, make sure you do all of the homework on that type of problem.

It's also important to know whether your teacher gives partial credit on a problem. Many—maybe most—math and science teachers do, because they consider taking steps toward a correct solution worthy of credit. Finally, you want to know on multiple choice or true-false tests whether you are going to be penalized for an incorrect answer. This is extremely important, because it determines whether you should guess.

If there is no penalty for guessing, you don't want to leave a single question unanswered. In fact, with three minutes to go and no penalty for guessing, you should fill in all of the unanswered questions. Some experts advise always guessing "false" on true-false questions and "C" on multiple choice questions. My advice is to guess randomly; that way, in the long run you should receive average credit.

Many tests, however, such as the SATs, have penalties for incorrect answers.[2] Expected value once again comes to your aid here, because you can use math to compute the expected value of a guess. For instance, if you have a five-answer multiple-choice question and a correct answer is worth 5 points, an unanswered question is worth 0 points, and a wrong answer is worth -2 points, your probability of getting the question correct is $1/5$ if it is a complete guess. As a result, the expected value of a complete guess is $1/5 \times 5 + 4/5 \times -2 = -3/5$. Because this result is less than 0, it is obviously best not to answer this question. If you can eliminate two of the five answers, though, you will now be guessing among three different answers. Your expected value in this case is $1/3 \times 5 + 2/3 \times -2 = 1/3$; because this result is greater than 0, it pays to guess. It's important to know this information going into an exam.

Let's assume you've got all of this information, and the instructor or the proctor passes out the tests and tells you to begin. What now?

It doesn't matter whether you are taking a multiple-choice test, a true-false test, or a problem test; look at the first problem and if you *immediately* know the answer or know how to do it, then do it. If not, go on to the next problem until you find one that it is easy and you can do it. If you can't find such a problem, you are either taking a graduate-level rocket science class or you didn't study enough. In the former case, you don't need this advice, and in the latter case, nothing can help you. You want to continue through this exam using the same strategy, so that on the first pass through, you've answered all of the easy questions.

There is a solid arithmetical reason for this. If you were planning on doing a job, would you rather take an easy one that pays $20 per hour or a difficult one that pays $10 per hour? The answer is obvious—and your instructor is paying you to answer questions, although he pays you in points per minute, rather than in dollars per hour. Picking the low-hanging fruit is the best way to start off an exam, for psychological reasons as well as logical ones. Most students are nervous at the start of an exam, and getting some points under your belt is a good way to alleviate this feeling.

After having gone through the exam once, what do you do next? If it's a true-false or multiple-choice exam, estimate the number of questions and the amount of time remaining, and compute the average amount of time you'll have for each question. Then go through the exam, allotting that average for each question. If you haven't figured out the answer by the allotted time, either guess or skip it, depending on whether the expected value of guessing is greater than 0, as discussed earlier in this section.

If you have a math or science exam with problems that have different point values, the analogue of this procedure is to figure out how many total points remain and how much time there is remaining, compute the average number of minutes per point, and then allot to each problem that average multiplied by the number of points for that problem. As an example, if 50 points remain to be completed and you have 30 minutes remaining, the average is 3/5 of a minute per point. Therefore a 15-point problem is entitled to $15 \times 3/5 = 9$ minutes' worth of work. Unless you can perform this computation easily and *really* quickly, my suggestion would be to work on the problem that has the smallest number of points, because it's probably the easiest. Don't dwell on it, though; if you're not getting anywhere, move on to the next problem. Dead time should be avoided at all costs.

Over the years, I've watched a lot of students take a lot of tests. I've seen some really good students and some really bad students, and I've not only observed them taking tests, but I've asked them questions. Even the better students often don't seem to realize that taking tests is a competitive endeavor, and competitive endeavors merit a look at strategy. My feeling is that while good test-taking strategy cannot make the difference between a C and an A, it can make the difference between a C and a B, or a B and an A.

In fact, I've seen many students do precisely the wrong thing during an exam: work furiously at the most difficult problem. I think there's a subconscious reason that students do this; they feel that if they get the hard stuff out of the way, it will be downhill sledding from there on in. True enough, but they may spend far too much time beating their heads against a stone wall. You can't afford to do this on an exam. Interestingly enough, getting the hard stuff out of the way is an excellent strategy for doing either routine tasks, such as washing the dishes or doing the laundry, or tasks that have a degree of interdependence, such as when a group of people has a lot of different tasks to do. Getting the harder or longer tasks out of the way in the latter instance makes bottlenecks—situations where you have one critical resource that several people need simultaneously—less likely.

Strategies for Improving Your GPA

Obviously, the first and most important step in improving your GPA is to improve your test scores—so if you skipped the previous section, go back and read it. Yet there are things you can do to boost your GPA beyond taking tests.

The first thing you can do is learn how to study effectively. I already mentioned that you shouldn't study three hours in a row unless the subject is something you really, truly love. Effective studying, however, doesn't necessarily mean the best way to memorize facts or learn procedures. One way to improve your GPA is to be aware of the mathematics of GPA computation.

Many students don't pay much attention to how their GPA is computed. Every so often, an inconsistency might appear to a student, usually something like getting an A, two Bs, and a C, and having a GPA under 3. Even if students take the time to unearth the reason for this inconsistency, they don't always use it to their advantage.

The reason for a GPA under 3 with an A, two Bs, and a C is that the courses are unequally weighted. At most schools, the GPA is computed by weighting the grade according to the number of units allotted to the course. Let's take a look at such a GPA.

Course	Units	Grades	Grade Points
Algebra II	3	B	9
Spanish II	3	B	9
U.S. History	3	A	12
Biology	4	C	8

Total units = 13

Total grade points = 38

GPA = 2.92

Grade points for each course are computed by multiplying the number of units by the numerical value of the grade (A = 4, B = 3, and so on), and the GPA is computed by dividing the number of grade points by the number of units. Biology is a lab course, sometimes requiring you to do icky stuff like dissect frogs, and lab courses are usually more heavily weighted than are courses that do not require a laboratory. Traditionally easy courses, from a grading standpoint, such as PE and music appreciation, are usually less heavily weighted.

How should this affect your studying? Most students put in a certain amount of baseline work in each course. Of course, this varies from course to course—some students are mathematically talented and don't have to work so hard to get A's in algebra— but it's a reasonable assumption that they put in roughly the same amount of time in each course. In case there's extra time available, however, and you can't decide how to use it, close calls should always go to the course with more units. It's worth 3 grade points to boost that B in Algebra II up to an A, but it's worth 4 grade points to boost that C in Biology up to a B.

Just as it's important to know how individual tests are being scored, it's important to know how your GPA is being scored. There is a significant difference between GPA strategy in schools that have standard grade awards of A = 4, B = 3, and so on, and schools that have plus/minus grades, in which a B+ = 3.3 and an A− = 3.7. When a school has plus/minus scoring, there is generally not much point in putting more effort into boosting one grade at the expense of another. Yet with standard grade awards (no plus/minus grades), your studying strategy can have a significant impact on your GPA. To take advantage of this opportunity, you must know where you stand prior to the final exam (or final paper) that occurs in almost all courses.

What you want to do here is play offense when your grade is in the high range (B+, C+) and play defense when your grade is in the low range (A−, B−). The reward for boosting a B+ to an A− is measurable, but there is absolutely nothing to be gained by boosting a B− to a B or even a B+. I score all exams on a scale of 0 to 100 (this is easy to do for math exams or multiple-choice exams but probably not so easy to do for essays), and I always let students know what score they need on the final exam in order to receive a particular grade. In my school, there is a rule that the final exam must count for between 25 and 33 percent of the final grade (the exact percentage is at the instructor's discretion), so if a student has a grade in the middle of a grade range (a B rather than a B− or a B+), it is not likely that the final exam score will change the final grade. A student who is averaging a solid B on the midterm exams probably cannot get an A in the course unless he or she turns in a perfect paper, and with a solid-B student that's not too likely to happen. On the other hand, unless the student gets a D+ or a C− on the final, he or she is almost certainly going to hang on to that B. So what's the point in killing yourself trying to improve the B to an A, when you're almost certainly doomed to failure? It's wasted effort.

The flip side of this particular coin is that you want to study especially hard to boost a B+ to an A− and to prevent a B− from falling to a C+. In fact, before you even open the books for the final, it's important that you take stock of exactly where you stand in your courses. Your studying strategy is very different when you have three solid Bs, as opposed to when you have one solid B, one B−, and one B+. In the first case, you probably don't have to do that much work to preserve your three Bs. In the second, you should study only enough in the course with the solid B to preserve the grade and devote the remainder of your effort to hanging on to the B− and boosting the B+ to an A−.

Some teachers are like me; they will let students know exactly where they stand prior to the final exam. Other teachers will expect the students to work it out on their own from whatever information they have (their test scores and the grading information that the teacher has supplied), but almost all teachers will give a straight answer if a student comes in to the teacher's office and asks directly how well he or she needs to do on the final exam to get a certain grade.

A student's time and effort are limited resources, and an individual wants to get the most value for the time and effort he or she invests. The topics discussed here are certainly not rocket science, but students are often unaware of them. I've seen students who are almost certainly destined for a B if their final exam lies in a 30-point range besiege my office for help during the last week of class. I give them the help they ask for, and they end up getting a B+ in the class—which translates to a B on the grade sheet. I hope they didn't forgo the opportunity to boost a B+ to an A− or to prevent a B− from slipping to a C+ in the process.

Some Grades Are More Equal Than Others

As Napoleon the pig so famously declared in George Orwell's classic *Animal Farm*, all animals are equal, but some of them are more equal than others. The same goes for grades and for GPAs.

In college, and to a lesser extent in high school, there is a difference between the student who gets high grades in a particular subject or group of subjects and the student who gets generally good grades. Many students come to college planning to pursue a career that requires an advanced degree. In many instances, the graduate school will consider the student's performance only in courses related to that particular degree. I was the graduate adviser in the Mathematics Department for a number of years, and we looked only at upper-division grades in math. In fact, there is no overall GPA requirement for admission to the graduate program in mathematics at my school; the only thing that is required is either a degree in mathematics or a degree with eight upper-division courses in mathematics (to accommodate the occasional major in physics, computer science, or economics who has minored in mathematics).

The other side of the coin is that some schools do not require a major in a particular subject. I just took a look at the FAQ section on the UCLA School of Law Web site, and one of the FAQs was, "What major do I need to be accepted to UCLA School of Law?"[3] The answer was that the UCLA School of Law recommends no particular major. If you are applying from my school, take note of the fact that math majors generally graduate with GPAs that are between .70 and 1.00 points lower than the GPAs of students who major in criminal justice (what do they do, sit around watching episodes of *CSI: Miami*?). If I graduated from high school and wanted to become a lawyer, the best way to achieve this is by gaining admission to a good law school, and I'd certainly think twice about becoming a math major as opposed to a criminal justice major. I'm sure the members of the admissions committee for the law school weren't born yesterday and realize that mathematics is probably a more demanding major than film or criminal justice is, but if the committee is considering two students with roughly equivalent LSATs (and I can't see how a course in differential equations will help you score well on the LSAT), and one is a math major with a 2.60 average and the other a film major with a 3.50 average, it might be hard to argue for the math major.

The Guessay Question

Many of my nonmathematical courses (I actually did take some, not only in high school but in college) featured what I used to call a "guessay question." The teacher would announce that an exam would have an essay question on one of two different topics, and it was a guess which topic he would choose when we actually took the exam. In general, I always felt that I could nail an essay topic if I had the time to study for it, but if I didn't study I'd have to BS. For some reason, the teacher always seemed to make me choose between a relatively easy topic, on which I could do a reasonable job if I had to BS, and a hard topic on which I had almost no shot without studying. Here's the problem: if you have time for only one question, which should you study?

Here's yet another opportunity to use mathematics to help your grades. First, you should make an estimate of how you think you would do in each of the possible four scenarios. You have two different topics you could study for, and the teacher has two different topics from which to choose the essay. I'm going to write down a game matrix to illustrate the basic idea, but there's a little more to this than simply applying game theory.

	Teacher Chooses	
	Easy Topic	Hard Topic
You Study		
Easy topic	A	D
Hard topic	C	B

Your first decision is whether you can afford to get a D or not. In college, some courses are crucial requirements for either the major or graduation, and you must get at least a C; others sometimes simply require a D to pass the course and get on with your life. If you can't afford to get a D, you obviously *must* study the hard topic.

It's remotely possible that you just had a feeling of déjà vu. Yes, you have seen a situation similar to this: in the first chapter when you took time out for an appearance on a quiz show. After you won $100,000, the emcee offered you the choice of keeping it or exchanging it for a presumably improved chance at winning $1,000,000. If you need the $100,000 for an operation for your child, it doesn't matter what the expected value of switching is. It's the same type of situation here.

Let's suppose, however, that it's simply a matter of GPA; you want to get your highest GPA in the long run. This puts the problem squarely under the game theory tent. Translating the letter grades to GPA numbers, the matrix looks like this:

	Teacher Chooses	
	Easy Topic	Hard Topic
You Study		
Easy topic	4	1
Hard topic	2	3

You can easily see that there's no pure strategy for you; you can't simply study for one of the topics and ensure that no matter which topic the teacher chooses, you will get better grades than had you studied for the other topic. Even though the teacher is not your opponent, the game is analyzed from the standpoint that he is, and the teacher cannot ensure a lower grade for you, no matter what you study, by choosing one topic rather than the other. So it's time to crunch the numbers.

	Teacher Chooses		
	Easy Topic	Hard Topic	ORD
You Study			
Easy topic	4	1	1
Hard topic	2	3	3

The answer? You should study the easy topic once and the hard topic three times. You can check that this gives you an expected GPA of 2.5, no matter which topic the teacher chooses.

I deliberately chose a matrix that required a mixed strategy, but this procedure is a good one to adopt when your time is limited and you have to choose which topic to study for a "guessay question." Incidentally, another possibility, adopted by many students, is to split their time between studying the two topics. Students will generally split their time 50–50 between the two topics, but assuming you make the same amount of headway per hour of studying each, you should split your time 75–25 in favor of studying the hard topic.

6

How Math Can Extend Your Life Expectancy

How dangerous is it to speed?

• • •

Why might your prescription show the wrong dosage?

• • •

Should you have a risky surgery or not?

I hope that before the copyright on this book expires my agent is able to sell its rights on the planet Vulcan (of *Star Trek* fame). After all, the way Vulcans greet one another is by uttering the phrase "Live long and prosper," and using arithmetic to achieve those goals is a primary focus of this book. This chapter addresses the first half of that greeting.

It's Not Just a Number

One of the first things we learn to do with mathematics is to measure things. Comparison involves measurement, and although it seems both obvious and trivial to mention it, numbers are often a measure of risk. As such, it behooves us to pay attention to them.

Many of these numbers are connected with our health. I can remember that when I was young, I was exposed to a startling fact: every extra pound of fat on the human body contains three extra miles of blood vessels through which the heart must pump blood. The heart is a very impressive muscle, but my guess is that it is capable of doing only a certain amount of work, and when that amount of work has been performed, the heart has had it. As a result, I've always paid pretty close attention to my weight. Similarly, stage 2 hypertension is characterized by a blood pressure reading of 160 or more (for the higher number), and that greatly increases the chances of having a heart attack, kidney failure, or a stroke. As the poet Andrew Marvell so sagely put it, "The grave's a fine and private place, but none, I think, do there embrace."[1] I'd like to do a lot more embracing, both literally and physically, before I depart, and so I'll take measures to keep my blood pressure down.

When the surgeon general's report connecting smoking with lung cancer came out in the mid-1960s, I cut my two-pack-a-day habit down to zero, although it took me almost two years to do it. I'm a math teacher, and I believe very strongly in numbers. Of course, I try to see how thoroughly documented the numbers are, but it goes against the grain for me to ignore numbers. Some numbers are simply numbers, but other numbers are not just numbers; they're warning signals, and we should pay attention to them.

Numbers also provide a means of comparison that is unmatched in its ability to cut to the chase. Although I have a certain amount of confidence in a Harvard-trained physician, as opposed to a physician's assistant or a physician trained at an institute of lesser prestige, what I really want to see on a physician (or a lawyer or an auto mechanic) is how many times this person has done procedures similar to what he or she is about to do on me (or for my legal case or on my car), and how many times this person has been successful. While I'd like to see legislation requiring this information to be made available, I don't really expect it to happen anytime soon. There is a relatively recent article titled "Physicians' Credentials: How Can I Check Them?" by Dr. Stephen Barrett, that can be found by searching the Internet. By the time this book is published, more complete information may be available.[2]

Ask yourself which you would prefer: a physician's educational background and career history, or a reliable database giving the equivalent of that professional's batting averages and slugging percentages. For me, it's not even close; I want the numbers.

Knowing that numbers can help keep us safe is important, but that's not really what this chapter is about. Most of us think of math as doing things with numbers that are a little deeper than simply comparing them. You know, addition, multiplication, division—all of that good stuff. So let's take a look at how math, the type that involves arithmetic operations, can help keep us safe in everyday life.

Speed Really Does Kill

The world of the early twenty-first century is far more hectic than the world of a century ago. Just look at what has happened to communication. A hundred years ago, a letter would take a week or so to go from coast to coast, and even into the middle of the century a telephone call from New York to Los Angeles cost $2 (and that's in 1950 dollars) for three minutes. Nowadays, many of us don't even wait to get home to read our e-mails; we pick them up off our PDAs and cell phones.

Unfortunately, this desire to speed things up can have catastrophic consequences when we hit the road. Even the proverbial little old lady now drives too fast, thanks to the seductive ability of modern automobiles to make speeds in excess of 75 miles per hour seem relatively slow. I remember when I was young that everything on the highway appeared to be rushing past while I was going 50; now everything seems leisurely when I go 70.

Let's look at how little you have to gain and how much you have to lose by increasing your speed from 60 to 70. My guess is that I live farther from work than most people do; I'm on the freeway for about 15 miles. Because 60 miles per hour is 1 mile per minute, the 15-mile journey would take me 15 minutes at that speed. At 70 miles per hour, we cover 1 mile in 6/7 of a minute, so the 15 miles are covered in $15 \times 6/7 = 90/7$ minutes, which is a little less than 13 minutes. I would save slightly more than 2 minutes by increasing my speed from 60 to 70. What price do I pay for this savings?

First, I have less time to react to potential danger. When I took driver education, the rule of thumb was that you should leave one car length's distance between you and the car in front of you for every 10 miles per hour of velocity. That rule of thumb might still be followed in less populated areas, but even when traffic on L.A. freeways is moving smoothly at high speed, there is very little difference between the actual gap between cars and tailgating. Even if you happen to be following the "one car length per 10 miles per hour" rule to the letter, that extra car length that you leave is probably not enough to prevent an accident. Remember, 1 mile per hour is 5,280 feet in 3,600 seconds, so in 1 second a car traveling at 1 mile per hour will travel 5,280/3,600 feet, approximately 1.5 feet. Your car is approximately 15 feet long. At 60 miles per hour, you are traveling about 90 feet per second, so six car lengths are covered in 1 second. This means that in case the car in front of you stops for one reason or another, you have a total of 1 second to react and for your brakes to stop you. If you are moving at 70 miles per hour, you are traveling about 105 feet per second, so the seven car lengths are again traversed in 1 second. Once again, you have a total of 1 second to react and for your brakes to stop you—but your brakes have to work a lot harder at 70 than they do at 60. You will have somewhat more time if the car in front of you merely slows down rapidly, but worst-case scenarios do happen. It goes without saying (but I'll say it anyway) that your margin is less if the visibility isn't so good or driving conditions make braking a more uncertain affair. I love L.A., but I'm amazed at how L.A. drivers seem to think that the idea is to drive faster in rainy weather. It seldom rains in southern California, but when it does I tend to avoid driving on the freeway.

If you happen to suffer an accident while driving at 60 mph, it's not going to be good—but it will be considerably worse at 70 mph. Kinetic energy, the energy associated with motion, increases as the square of the velocity, so the ratio in the energy of a car traveling at 70 mph to that of one traveling at 60 is $70^2/60^2 = 1.36$. Thus, a car traveling at 70 mph has over a third more kinetic energy than one traveling at 60 mph. At 75 mph, a car has more than 50 percent more kinetic energy than one traveling at 60 mph. Princess Diana's limo didn't survive a crash in which the car's speed was estimated to be around 60 mph— and neither did Princess Di. Are you really sure you want to save those 2 minutes on a 15-mile trip?

I'm not the first person in the sciences to express interest in this topic. Max Tegmark, a physicist at MIT, has actually done some expected-value calculations based on data compiled in the early years of the twenty-first century. His conclusions are worth listing—and considering. His calculations are easy to follow, certainly for readers of this book.[3]

Each hour of driving on an interstate freeway decreases life expectancy by 19 minutes. That's a stunner. I probably spend an average of 6 hours a week driving on interstate freeways, which means that my drive time each week cuts my life expectancy by about 2 hours. That's about 4½ days a year, and during the course of a lifetime, maybe 8 months of my life expectancy are lost to driving on the interstate freeway. It's worth it to me, though, especially because I know they'll take it off the end of my life and not the middle.

Each hour of driving in local city traffic decreases life expectancy by 8 minutes. It's obviously safer to drive in the city; the speed limits are lower. I wonder if Tegmark has included in his calculations the chance of being carjacked.

Each hour spent riding a motorbike decreases life expectancy by 5 hours. This ought to discourage the motorcyclists who buzz by with abandon at 80-plus mph on the freeway. Frankly, even before I read Tegmark's statistics, I would never have taken the risk. You simply cannot afford to have an accident on a motorbike—and accidents are inevitable.

Each domestic U.S. flight decreases life expectancy by 13 minutes. And that's why flying is safer than driving.

Percentages: The Most Misunderstood Topic in Mathematics

It's a continuing source of amazement to me how many errors are made involving percentages. I've read articles by economists at prestigious think tanks who commit the most glaring gaffes in calculations that deal with percentages. Percentages are so frequently screwed up that I decided to give one of my classes an informal survey.

Math 109 is a terminal math course, generally taken by students who are not planning to major in a subject that doesn't have specific math requirements. Nursing majors, for example, need to take a course in statistics, but history majors can simply take a course like Math 109 to fulfill the distributional mathematics requirement of the university. I teach a section of Math 109 for our school's Honors Program, which generally draws bright and inquiring students. I decided to find out something about their general level of mathematical knowledge and so administered a quiz with the following four questions. You might take a few minutes to try them. I'll answer them fairly quickly later.

1. The price of gas went down 10% last week. This week it went back to the price it was at the start of last week. By what percentage did the price of gas increase this week?

2. The price of gas went up 10% last week. This week it went back to the price it was at the start of last week. By what percentage did the price of gas decrease this week?

3. The price of gas goes up 10% this week and 10% next week. By what percentage does the price of gas rise during this two-week period?

4. The price of gas goes down 10% this week and 10% next week. By what percentage does the price of gas fall during this two-week period?

Fifteen students took the exam; out of 60 answers there were a total of 6 correct answers, 4 of which were supplied by Mara, the best student in the class. Ten of the papers were identical, answering 10% for questions 1 and 2 and 20% for questions 3 and 4.

The easiest way to do "pure" percentage problems, those in which only percentages are involved, rather than tangible units such as money or volume, is simply to operate from a base of 100. For problem 1, if you start with a gasoline price of 100 and the price drops 10%, since 10% of 100 is 10, the new price is $100 - 10 = 90$. In order to get back to the original price of 100, the price must go up 10 units from a base of 90, a percentage increase of $100 \times 10/90 = 11\frac{1}{9}\%$. A similar reasoning shows that in problem 2 the price must fall 10 units from a base of 110, a percentage decrease of $100 \times 10/110 = 9\frac{1}{11}\%$. Incidentally, when students say the price of gasoline is $2.00 a gallon, not 100, I tell them that the price really is 100—it's just that it's 100 two-cent pieces, and that the price of anything is always 100. It's simply a matter of figuring out what the units are, and this can be done by dividing the actual price by 100.

The other two problems are handled similarly. In problem number 3, if one starts with a price of 100, the first week it goes up 10% of 100 to 110 (just as in problem 2), but the next week it goes up 10% of 110 to 121. In the two-week period, it has gone up 21 units on a base of 100, which is 21% (you could either recognize this or do it by the percentage formula of $100 \times 21/100 = 21$). On problem number 4, if one starts with a price of 100, the first week it goes down 10% from 100 to 90, and the next week it goes down 10% from 90 to 81, so it has gone down 19%—from 100 to 81—in two weeks.

Recall that 10 out of 15 students in an honors class made the same mistake on every problem. They failed to take into account that when computing percentages, the percentage is always computed on the *current* base, not on the *previous* base. A 10% drop in price, followed by a 10% rise in price, does not get you back to the original price because the 10% drop is computed using the original price as base, whereas the 10% rise is computed using the price after the drop as base. Because this number is lower than the original price, the 10% rise in price from the lower base cannot offset the 10% drop from the original (higher) base.

It's not only my students who are prone to making errors in percentage calculations. On November 5, 2008, the day after the presidential election, I heard a noted conservative radio commentator who shall be nameless (hint: he graduated from the same esteemed institution that I did) cite the following numbers: In 2004, 37 percent of voters identified themselves as Republicans, 37 percent as Democrats, and the remainder as independents. In 2008, 39 percent of voters identified themselves as Democrats, 32 percent as Republicans, and the remainder as independents. After quoting these numbers, the commentator noted that 5 percent of Republicans had switched to becoming either Democrats or independents. How many errors can you spot in that statement?

You receive full credit if you caught these two: First, the voter base was different in 2008 than it was in 2004, and because the numbers were different, one cannot reach any legitimate numerical conclusion. It's conceivable that not only were the number of voters different in 2008 from 2004, but the voters themselves were different; many who voted in 2004 might not have voted in 2008, and vice versa. Second, and relevant to the type of thing we have been discussing, even if the voter base were identical, 5 percent of *voters* switched from Republican to Democrat or independent. If you work with a base of 100, however, 5 of 37 *Republicans* switched affiliation, and that's about 13.5 percent.

So Many Ways to Die

The misunderstanding concerning percentages can kill you in a number of different venues. The one that is most obvious to me occurs with regard to drug dosages. Drug dosages by themselves are sometimes misunderstood; a recent high-profile situation arose at the prestigious Cedars-Sinai medical center in Los Angeles when twins born to the actor Dennis Quaid were given 1,000 times the prescribed amount of the blood thinner Heparin.[4] This type of situation happens with such frequency that it has even been given the name "death by decimal point." It occurs when a pharmacist cannot tell where the doctor has placed the decimal point, or when someone erroneously fills a prescription in which the concentration of the drug is given on the bottle in milligrams per milliliter when the doctor actually prescribed the drug in a concentration of milligrams per liter. This isn't even subtle, yet it involves mathematics, at least to some extent.

Death by percentages is mathematically more sophisticated—yet just as fatal. Because it is easy to describe changes in medication using either percentages (or fractions, where many of the same problems arise), the "change of base" error can have far-reaching consequences. I'd always thought that well-educated people were comfortable with percentages, until I read an op-ed piece in the early 1990s by an economist at the Hoover Institute in Palo Alto, which included a major error on percentages.[5] It occurred to me that similar errors could happen with regard to medical prescriptions. A patient on a life-saving medication might be doing well, and the doctor could request that the dosage be cut by 75 percent. A relapse occurs, and the doctor either makes the percentage error by requesting that the dosage be raised by 75 percent, or the doctor informs the person responsible for providing the medication that the dosage should be restored to its original level—whereupon the dosage is increased by 75 percent. As you can see, if you start from a base of 100, cutting the dosage by 75 percent reduces it to 25, and increasing it by 75 percent raises it to 43.75, less than half the original dosage.

The "change of base" error can also result in overdoses. There are actually two different ways that the overdose can result: from a failure to calculate accurately or a failure to communicate accurately. One possibility is that a doctor might raise a patient's dosage by 100 percent and then do so again, feeling that he was tripling the original dosage when in reality he was quadrupling it.

Of course, errors like this shouldn't occur—but then they shouldn't have lost that multimillion-dollar Mars orbiter either, because some team members were using English measurements and the others were using metric measurements.[6] As I mentioned, I've seen PhDs in economics, a subject heavy on math, mess up percentages, so I was happy to learn that doctors are instructed to write medical prescriptions in specific amounts, to ensure that no "change of base" errors occur. Nowadays, however, not all medication decisions are made by doctors, so I hope that all medical personnel, from doctors to nurses to paramedics, stick with the practice of giving specific amounts for all medications.

May You Never Have to Use This

Sometimes math can hit really close to home. Such a situation happened to me in the late 1980s, when my father went into the hospital for what seemed like an endless series of visits. He was a strong and stoic man who had fought his way through a heart attack, a stroke, and intestinal problems. I had medical power of attorney for him while he was in the hospital, and when he was recovering from a procedure I received a call from the surgeon. There was a new and somewhat risky procedure that the surgeon wanted me to authorize. I talked to him for about ten minutes and felt that he was sincere about wanting to help my father, rather than just wanting to do the procedure merely for the sake of trying something new. Although I couldn't pin the doctor down to numbers, it was clear that my father either had a specific condition that was likely to be helped by surgery or did not have the condition that the procedure was designed for.

I asked a doctor I knew for guidance, but it wasn't his area and he wasn't familiar with the procedure, although he did look it up in a journal. The procedure was described as new and promising—and risky. I made the following rough estimates for my father's survival chances after talking with both the surgeon and the doctor.

	Father Needs Surgery?		
	Yes	No	ORD
Try Surgery?			
Yes	60	30	70
No	10	80	30

Needless to say, there was no pure strategy available. I suspect that in similar situations no pure strategy is ever available. The odds were therefore 7 to 3 in favor of surgery, and by checking this against the case in which Dad needs surgery, you can see that his chances of survival were $(7 \times 60 + 3 \times 10)/10 = 45\%$.

As you might imagine, my faith in game theory was not so overwhelming that I was ready to risk my father's chances for survival on a randomizing device, so I desperately tried to think of an alternative. If I could not come up with one, I felt fairly certain that my father would have understood what I was trying to do—when he was well into his seventies, he still spent Sunday mornings listening to lectures delivered on mathematics by a professor at the University of Chicago. Fortunately, I didn't have to make this decision or explain it to my father. When he became conscious, he took the decision out of my hands by telling me that he simply didn't want to undergo any more invasive procedures. I respected his decision, as I hope my wife will respect mine if it ever comes to that. It was the last significant decision that he made, because he died a week later. For perhaps the only time in my life, I was grateful not to have the opportunity to apply my knowledge of mathematics.

7

How Math Can Help You Win Arguments

Was the bailout the only way to save the banks?

• • •

Do you really have logic on your side?

• • •

What are the first arithmetic tables learned by children on Spock's home planet?

When I was in school, I took courses in both science and philosophy, feeling that science was the search for knowledge and philosophy the search for wisdom. Logic underpins both, and as Commander Spock of *Star Trek* once put it, logic is the beginning of wisdom, not the end—but on the road to wisdom, you've got to start somewhere.

The $700 Billion Question

My, how the stakes have been raised since I was a kid. I remember listening to *The $64 Question* on the radio and being amazed when the jackpot escalated by three orders of magnitude on television's *The $64,000 Question*. Fast forward a few decades, and welcome to *Who Wants to Be a Millionaire?* But all of that pales in comparison to the recent attempt to convince the American people of the logic of bailing out the investment banks to the tune of $700 billion—give or take a trillion. The argument goes something like this:

1. If we do not loan $700 billion to the banks, the credit market will freeze up.
2. If the credit market freezes up, the economy will be greatly damaged.
3. Therefore, if we loan $700 billion to the banks, the economy will not be greatly damaged.

This is basically the template for the argument used by every industry seeking regulation that will be favorable to it. But is it a logical one?

The Ultimate Trump Card

One of the most common methods of winning an argument is to declare that you have logic on your side. Logic is recognized by almost everyone to be the ultimate trump card; practically no one challenges an argument that is acknowledged to be logical. The rules of combat allow one to challenge the conclusion of an argument that is acknowledged to be logical by challenging the premises of that argument, but few people tend to challenge the logic of the argument. One possible reason is that most people have not made any sort of a systematic study of logic. They recognize some simple arguments as logical, and they can often recognize glaring errors such as ad hominem appeals, but the general understanding of logic is as woefully lacking as the general understanding of mathematics.

Here's a relatively simple example. A common expression is "If the shoe fits, wear it." Many people believe that the statement "If the shoe doesn't fit, don't wear it," follows logically from that one. After all, if the shoe doesn't fit, why on Earth would you want to incur blisters, calluses, and possibly foot problems by wearing it? This particular injunction, not to wear shoes that don't fit, is accepted by virtually everyone (at least, everyone who wants to avoid blisters, calluses, and foot problems), but it is not a logical conclusion, as you shall see in the rest of this chapter when we study symbolic logic.

Symbolic Logic

I often teach Math for Liberal Arts Students, aka Math for Poets and other less flattering names. Usually, the students in this course approach the first lecture with only slightly less apprehension than they do a visit to the dentist. They know it's going to hurt, they're just not sure how much. I always start these classes off with symbolic logic—because it doesn't hurt a bit, even for poets.

There are a lot of different ways to combine two numbers. You can add them, subtract them, multiply them, divide them, exponentiate them, or take the larger (or the smaller) of the two numbers—and that covers only the common ways of combining them. Yet school always starts with the addition and multiplication tables. My guess is that these were regarded as basic for commerce: you need addition to total up the cost when someone buys a lot of different things, and you need multiplication (which is the repeated addition of the same number) when someone buys a lot of identical things that cost the same price.

People learn arithmetic not because arithmetic intrinsically has a purpose but because a need exists that arithmetic can satisfy: the simple need in commerce to compute the bill. If people need to learn how to determine whether an argument is logical, they can also learn arithmetic—only it's not quite the same arithmetic, and the tables are a whole lot simpler. For the most part, you already know them.

The arithmetic of logic was constructed to deal with statements that are either true or false. To adapt this arithmetic for a computer, statements that are true are assigned the value 1, and statements that are false are assigned the value 0. Just as numbers apply to some things but not to others, the labels "true" and "false" apply to some statements but not to others. By *true* and *false*, we mean statements that are universally accepted as one or the other, for whatever reason: "$2 + 2 = 4$" is a true statement for arithmetical reasons, "Fresno is the capital of California" is a false statement because Sacramento is the capital of California, and "Fresno is a great place to live" is neither true nor false because it's a matter of opinion. Just as we could use the label "one" instead of the digit "1" when we do ordinary arithmetic, we could use the word *true* or the digit "1"—or we could adopt a halfway position and use T for true and F for false. I'll go that route, because it's sort of a halfway measure, preserving the single-symbol advantage of digits, yet reminding us of what the symbols being used represent.

Now, how do we want to work with this new variety of numbers? Fortunately, the English language (and most other tongues) already has a number of ways of producing new statements from existing ones. The most common ways are negation, conjunction, disjunction, and implication—which you know by the words most often used to accomplish them: *not, and, or, implies.*

Negation

The negation of a statement can be obtained in one of two ways. The simplest is by judicious positioning of the word *not*. If p is the statement "Fresno is the capital of California," then *not p* is the statement "Fresno is not the capital of California." In case it's hard to figure out where to place the "not," simply stick "It is false that" in front of the statement, obtaining (in this case) "It is false that Fresno is the capital of California."

The arithmetic table for *not* is quite simple. We'll use the letter p to represent a statement.

p	*not p*
T	F
F	T

The first row of the table says that when p is true, *not p* is false, and the second row says that when p is false, *not p* is true. You didn't have to learn a whole lot there, did you?

Evaluation of a statement is simply a matter of working from the inner parentheses first, just as you would evaluate an arithmetic expression such as $(2 + 3 \times (4 + 5))$. This produces the following arithmetical statements:

$$(2 + 3 \times (4 + 5))$$

$$(2 + 3 \times 9)$$

Because too many parentheses are a giant pain in the you-know-where, mathematicians have developed a hierarchy of operations known as PEMDAS (memorized by generations of schoolchildren as "Please Excuse My Dear Aunt Sally"). This mnemonic represents the precedence of operations in evaluating an expression such as $2 + 3 \times 9$: Parentheses, Exponents, Multiplication, Division, Addition, Subtraction. M appears before A in PEMDAS, so one performs the multiplication first. Continuing,

$$(2 + 27)$$

$$29$$

If you knew that p were true, you could evaluate the truth or falsity of *not* (*not* p) simply by working from the inside out as above.

not (*not* T)

not F

T

Do the same thing for a false statement p and you would find that *not* (*not* F) is evaluated to F. In other words, *not* (*not* p) and p always have the same truth value. Arithmetic uses the equal sign to indicate that two expressions have the same numerical value, no matter what the values of the variables making up the expression: $x + x = 2x$. This is called an identity, and some texts will use a three-bar equal sign to denote an identity: $x + x \equiv 2x$. In logic, when two expressions such as *not* (*not* p) and p have the same truth value, no matter what the truth values of the statements making up the expressions, such as p and *not* (*not* p), we say they are logically equivalent.

Conjunction and Disjunction

The conjunction operator joins together two statements p and q by means of the word *and*. In accordance with common usage, p *and* q is true only when both of the statements p and q are true. "Sacramento is the capital of California" and "$2 + 2 = 4$" are true statements, and everyone would agree that "Sacramento is the capital of California and $2 + 2 = 4$" is a true statement, although most people would wonder why you decided to combine those two. Similarly, "Los Angeles is the capital of California and $2 + 2 = 4$" would be judged to be false; all it takes is for one rotten (false) apple to spoil the conjunction barrel.

The disjunction operator joins together two statements p and q by means of the word *or*, but there is an ambiguity here. We use the word *or* in two different ways in English. The exclusive "or" requires us to choose exactly one alternative, such as "Did you vote for McCain or Obama?" The inclusive "or" allows us to choose both alternatives. When your server asks, "Would you like coffee or dessert?" he or she will be delighted if you select both because the cost of the meal, and therefore the tip, is bound to increase. Symbolic logic has adopted the inclusive "or," and so the statement *p or q* is false only when both p and q are separately false.

As a result, we have the following arithmetic tables for logic, which are commonly referred to as truth tables.

p	q	p and q	p or q
T	T	T	T
T	F	F	T
F	T	F	T
F	F	F	F

It's a lot easier to memorize truth tables than it is addition and multiplication tables. First, you already are familiar with the underlying structure of the English language. Second, multiplication tables have $9 \times 9 = 81$ entries (even worse if you have to memorize the times tables for 10, 11, and 12), whereas truth tables only have $2 \times 2 = 4$ entries for each operation.

Implication

The heart and soul of symbolic logic, and its raison d'être, is the implication operator *p implies q* (written in many books as *if p then q*). Although there is a certain value to discovering that *not* (*p or q*) is logically equivalent to (*not p*) *and* (*not q*), you knew that anyway: when your server asks you whether you would like coffee or dessert and you answer no, both parties are aware that this is equivalent to your not wanting coffee and your not wanting dessert.

The purpose of symbolic logic is to determine when an argument is valid by highlighting the only instance in which an argument is guaranteed to be invalid: when it could proceed from a true premise to a false conclusion. As a result, the truth table for *p implies q* is purpose-directed; it is false only when *p* is true and *q* is false. Schematically,

p	*q*	*p implies q*
T	T	T
T	F	F
F	T	T
F	F	T

My liberal arts students have no difficulties with the first two rows. Well, that's not exactly true; they sometimes have difficulty with accepting the truth of the statement "The capital of California is Sacramento implies that $2 + 2 = 4$." I can certainly understand this; there is no logical connecting argument between an odd geographical factoid and a mathematical truth. The implication operator, however, is simply a fraud-detecting device, designed to ferret out the obviously erroneous arguments that start with a true premise *p* and end up with a false conclusion *q*. This also explains why the last two lines of the table are true implications. Implication was not designed to examine an argument to see whether it consists of a sequential progression of statements, each of which follows logically from preceding statements; it was designed to detect an argument that is clearly fraudulent.

Evaluation of a complicated compound statement proceeds, as does evaluation of a complicated algebraic expression, by working from the innermost parentheses outward. In the following example, we will assume that *p* and *q* are true statements and *r* and *s* are false ones; we simply evaluate the truth value of a complicated expression one step at a time.

p implies (r or (q and not s)
T implies (F or (T and not F)
T implies (F or (T and T)
T implies (F or T)
T implies T
T

Piece of cake.

With all of this heavy artillery, we are now prepared to wage the ultimate battle: determining whether an argument is valid—or not.

When Is an Implication Valid?

An implication is valid if it is true independent of the truth of the simple statements that comprise it. Let's look at a really simple argument that everyone would agree is valid a priori; (*p and q*) *implies p*. There are two ways we could demonstrate this. The most straightforward way is to construct a truth table that looks at all of the possible truth value combinations for both statements.

p	*q*	*p and q*	(*p and q*) *implies p*
T	T	T	T
T	F	F	T
F	T	F	T
F	F	F	T

Not so bad, but there is a shorter way (especially when there are more than two statements involved in the argument): try to "falsify" the implication by giving it a false premise and a true conclusion. It can't be done here; in order for the conclusion to be false, *p* (the conclusion) must be false. This ensures that *p and q* will be false, but then

(*p and q*) *implies p* reduces to *F implies F*, which is *T*.

Validating Arguments

Part of the difficulty people have with deciding whether an argument is logical is that they do not distinguish between the form of the argument (which uses only letters to represent statements) and the specific argument presented (which uses actual statements). In order for an argument to be logical, the form must be such that you are *never* led astray; it is impossible to assign truth values to the individual statements that result in a true hypothesis and a false conclusion. An argument can be correct in a specific instance but not be valid. It is similar to the difference between an equation and an identity in simple algebra. The equation $x + 2 = 5$ is correct only when $x = 3$, but the identity $x + x = 2x$ is valid for all values of x. Numerical values in algebra are the equivalent of truth value assignments in symbolic logic.

Now would be a good time to return to the argument we examined earlier in the chapter. Is "If the shoe doesn't fit, don't wear it" a logical conclusion from the hypothesis "If the shoe fits, wear it"?

Let *p* denote the statement "The shoe fits" and *q* the statement "Wear it (the shoe)." The first statement, "If the shoe fits, wear it," is abbreviated as "*p implies q*," and the second statement, "If the shoe doesn't fit, don't wear it," as "*not p implies not q*." The argument is therefore

(*p implies q*) *implies* (*not p implies not q*).

Our mission is to determine whether we can "falsify the argument" by determining if there are truth values for p and q for which this implication is false. If that could be done, then p implies q would have to be true, and *not p* implies *not q* would have to be false. For *not p* implies *not q* to be false, *not p* must be true (that is p is false) and *not q* must be false (that is, q is true). But when p is false and q is true, p implies q is true, so the argument can be falsified.

Some people might claim that the only case to be considered in the argument is the case where the shoe fits, that is, p is true. Although the argument can be shown to be true in all cases when p is true, that is not the entire argument, and one must take into consideration the situations when p is false—just as one cannot claim that the statement $x + 2 = 5$ is true because one has found an instance, $x = 3$, for which the statement is true.

The mechanism we have developed is used not to reveal truth but to detect when the form of the argument renders it susceptible to the logical error of a true premise implying a false conclusion. Let's look at a typical argument one might have heard or read in the newspapers a few years ago:

1. If Bush listens to the military, the war in Iraq will not drag on after victory is declared.

2. Bush did not listen to the military.

3. Therefore the war in Iraq will drag on after victory is declared.

One might debate whether the first sentence is the result of 20–20 hindsight, but no one can doubt that both of the subsequent statements are true. So, is this argument logically valid? It starts from something that most people would accept in retrospect, and every other statement is true. How do we analyze this?

Remember that the purpose of symbolic logic is not to detect truth but to issue a warning when one can start with a true premise and end up with a false conclusion. There are just two basic statements in the argument. Let's abbreviate "Bush listens to the military" as p, and "the war in Iraq will drag on after victory is declared" as q. With these abbreviations, the first line becomes p implies *not* q, the second line is *not* p, and the third line is q. The argument uses the conjunction of the first two sentences as the hypothesis of an implication whose conclusion is the third sentence. Many "three-line arguments" have this form: a giant implication obtained by sticking "If" in front of the result of joining lines one and two together with an "and," concluding by putting "then" in front of the third line. Symbolically, the argument becomes

> *((p implies not q) and (not p)) implies q.*

The argument is valid if no matter what the individual truth values of p and q are, the big implication in the above line is always true, just as was the case in the previous argument. It's a little messy here to construct the truth table, so let's see if we can falsify it.

In order to falsify, the conclusion must be false, so the truth value of q is F. Since the hypothesis must be true, both of the statements (p *implies not* q) and *not* p must be true; since *not* p must be true, p must be false. Inserting these truth values into the argument and evaluating gives

> *((p implies not q) and (not p)) implies q*
> *((F implies not F) and (not F)) implies F*
> *((F implies T) and T) implies F*
> *(T and T) implies F*
> *T implies F*
> *F*

The argument about Bush and Iraq may be correct, in that the actual truth values for this particular argument (p and q

have both been shown by history to be false) constitute a true implication, but this is just one line of the four possible lines in the truth table corresponding to all possible truth value combinations for p and q. The argument is valid (aka logical) only if all four possible truth value combinations for p and q result in a true implication.

Analyzing the $700 Billion Question

It's finally time to take an analytical look at the argument with which this chapter began:

1. If we do not loan $700 billion to the banks, the credit market will freeze up.
2. If the credit market freezes up, the economy will be greatly damaged.
3. Therefore, if we loan $700 billion to the banks, the economy will not be greatly damaged.

You now have the tools you need. If you would like to try your hand at this, take a time out before reading the analysis. You have two choices: an eight-line truth table (ugh) or falsification. Let me suggest that falsification may be the easier route.

Let's start by abbreviating statements. Let p be "We loan $700 billion to the banks," let q be "The credit market will freeze up," and finally, let r be "The economy will be greatly damaged. The first line of the argument is "*not p implies q,*" the second line is "*q implies r,*" and the last line is "*p implies not r.*" Gluing the first two lines together as the hypothesis of a giant implication, of which the third line is the conclusion, gives us the statement

((not p implies q) and (q implies r)) implies (p implies not r).

In order to falsify this, the conclusion *p implies not r* must be false, which can happen only when *p* is true and *not r* is false—in other words, *r* is true. Because the hypothesis must be true, and the hypothesis consists of two statements joined together by *and* (nerdspeak: the two statements are "anded together"), each statement must be true. If *not p* is false, the statement *not p implies q* is true, no matter what the truth value of *q*. Similarly, if *r* is true, the statement *q implies r* is true, no matter what the truth value of *q*. Consequently, letting all three statements be true should falsify the argument. Let's see.

((not p implies q) and (q implies r)) implies (p implies not r)
((not T implies T) and (T implies T)) implies (T implies not T)
((F implies T) and (T implies T)) implies (T implies F)
(T and T) implies F
T implies F
F

It may be the winning move to throw $700 billion at the investment banks (I sure hope we get something for our money), but you'll never convince me that it's logical to do so, because the argument logically sucks.

In the book *It Takes a Pillage: Behind the Bailouts, Bonuses and Backroom Deals from Washington to Wall Street*, the former Goldman-Sachs money manager Nomi Prins argues that the actual figure is $12.7 trillion, not a measly $700 billion. I checked with the U.S. Bureau of Economic Analysis and found that the GDP for the United States in 2008 was about $14 trillion.[1] I certainly hope that the $12.7 trillion, if that's the true number, is amortized over a century or two, because I'd hate to think that roughly 13 of every 14 dollars that the United States produced last year went to the bailout.

It Really Is Arithmetic

You may be convinced of the value of logic, but you're a little skeptical about whether it belongs under the heading of "arithmetic." Let me try to convince you.

We're going to use just multiplication, subtraction, and one other operation: taking the larger of two numbers. This is usually written $\max(a,b)$ (for the maximum of a and b); $\max(5,3) = 5$. Oh, yes, to make life even simpler, we're going to use only the numbers 0 (which will correspond to F) and 1 (which will correspond to T). We'll let the letter p stand for a proposition, and the letter P for a number corresponding to the proposition p; when p is T, P is 1, and when p is F, P is 0. Now let's look at the truth tables and the operation tables together; I've put the truth tables in the format in which one usually sees addition and multiplication tables. Let's first compare the operations *not p* and $1 - P$.

		not p	
		T	F
p		F	T

		$1 - P$	
		1	0
P		0	1

You can see the obvious similarity, as you can in the following comparison of the tables for *p and q* and *PQ*.

p and q			
		q	
		T	F
	T	T	F
p	F	F	F

PQ		
	Q	
	1	0
P 1	1	0
0	0	0

The similarity doesn't stop here. Look at the tables for *p or q* and max(*P,Q*).

p or q		
	q	
	T	F
p T	T	T
F	T	F

max(P,Q)		
	Q	
	1	0
P 1	1	1
0	1	0

And the last piece of the puzzle, the tables for *p implies q* and max(1−*P,Q*).

p implies q		
	q	
	T	F
p T	T	F
F	T	T

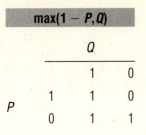

max(1 − P,Q)		
	Q	
	1	0
P 1	1	0
0	1	1

This phenomenon, the direct similarity between two mathematical systems, is known as isomorphism. The practical implication is that logic and arithmetic are merely two different ways of looking at the same idea. It may not be the arithmetic that we're familiar with, but it's arithmetic all the same. According to the Starfleet database, it's the first arithmetic tables that children learn on Vulcan, the home planet of the eminently logical Mr. Spock.

8

How Math Can Make You Rich

How can you actually make money off credit card companies?

• • •

Will refinancing your house actually save money?

• • •

Is a hybrid car a better value?

Even though supercomputers today are capable of carrying out trillions of computations per second in the quest to solve some of the really deep problems involving science, medicine, and engineering, the odds are that most of the arithmetic most of us do will be related to money.

Financing: You've Just Got to Do the Math

Most people don't look very deeply into financing. They should. It can seriously impact your quality of life. Although there is

107

nothing earthshaking or new in this chapter, from the standpoint of either mathematics or economics, it's nonetheless extremely valuable, and reading the chapter will almost certainly better prepare you to make financial decisions.

Borrowing Money: The Engine of Commerce

I'm not an economist, but I believe that there has been no single economic development that has had as much impact on the advance of civilization as the idea of paying money to borrow money. When you think about it, it's a natural idea: money buys goods and services; it performs work. Something that performs work should be paid for this service.

Most of our major purchases are paid for with borrowed money. Many of us could not go to college were it not for borrowed money, most of us could not buy a car if we had to pay cash, and almost none of us could own a house if we had to pay the full amount in order to live in the house. It is inconceivable that we would have developed the technology that so improves our lives without the financial infrastructure that enables people to own things by purchasing on credit.

Much of the mathematics of finance concerns future payments for things bought on credit. It centers on a very important idea: the present value of a payment.

The Indisputable Value of Present Value

Consider the way interest works. If the annual rate is 6% and you borrow $100 for one year, you must pay back $106 a year from now. The present value of a payment of $106 a year from now, when interest rates are 6%, is therefore $100; you need to stick $100 in a bank paying 6% interest *in the present* in order to have $106 a year from now. Yet if you need to have $106 a year from now and interest rates are 4.8%, you must obviously put more in the bank now—to be precise, you need to deposit

$101.15 now. We say that the present value of $106 a year from now at 4.8% is $101.15. If, however, interest rates have gone up, say to 7%, you need to deposit only $99.07 now to have $106 a year from now. When interest rates decline, the present value of a future debt increases—so you need more money *now* to be able to pay it off in the future. Conversely, when interest rates increase, the present value of a future debt decreases. The present value of a sequence of payments, such as the remaining monthly payments on your mortgage, is computed by simply adding up the present value of each of the payments.[1]

If you think about it, the way that present value is affected by the change in interest rates is really not very surprising. Suppose, for simplicity of computation, that interest rates are 5%, and you need to make a payment of $1,000 every year for a purchase you have made. If you have $20,000 in the bank now, at the end of a year you will have made $1,000 (5% of $20,000) in interest. You simply peel off the interest, pay the $1,000, and leave the $20,000 in the bank, where at the end of the next year you will have another $1,000 interest to make the next payment, and so on. If the interest rates decline to 4%, you will make only $800 in interest and will have to perform the dreaded "dip into capital" to make the payment. If, however, interest rates increase to 6%, you will make $1,200 in interest and can make the payment and have $200 left over—either to spend on wine, women, and song or to deposit in your account to earn extra interest.

Inflation and Interest Rates

Inflation is an increase in the cost of goods and services as time goes by. Often, this increase is driven by market forces—in the last two years, the price of a gallon of gasoline almost doubled (before it dropped) because demand had increased, the supply had not kept up with the demand, and the Middle East, the area responsible for much of the world's gasoline, remains highly unstable. In the same period, however, the price of an average home has declined. Various indicators have been devised to measure the overall inflation rate, loosely defined as the average increase of the cost in goods and services that would be purchased by a typical family.

Ideally, one would like to borrow money at a rate lower than inflation. For simplicity, imagine that you borrow $100,000 at 4% when the inflation rate is 5%. You could simply buy $100,000 worth of goods now, sell them for $105,000 in a year, spend the $104,000 required to pay back the loan, and have $1,000 left over for a rainy day. Of course, this example is highly unrealistic and oversimplified, but at least it indicates why it's a good thing to be able to borrow money at a rate lower than inflation. The flip side is that if you borrow money at a higher rate than inflation, there is the potential for trouble. In the previous example, if you borrow $100,000 at 4% when the inflation rate is 3%, you will be able to sell the goods for only $103,000, leaving you $1,000 short of the $104,000 that is needed to pay back the loan.

The greater the difference between the inflation rate and the interest rate, the larger the hole you will have to dig yourself out of. There are two major credit pitfalls that are part of the American scene: credit cards and home buying. Although libraries could be filled with books on these and related subjects, it's possible to cover most of the key ideas in a relatively short presentation.

Credit Cards: Their Utility and Some Associated Traps and Pitfalls

I teach on a college campus, and virtually every day I walk past a booth (or booths) where students are being induced to sign up for credit cards with offers of free T-shirts or iPods or God only knows what else. Many students look at credit cards as an instant passport to a better standard of living. They're not dumb; they realize that they will have to pay interest on the unpaid balance of the credit cards, but they also realize that they have very long lives ahead of them and will undoubtedly make a whole lot of money. What's to worry about?

It's not just students, however, who get such offers. Check your mailbox and your inbox. The chances are pretty good that you, too, are deluged with offers for credit cards, only if you're not a student the inducements are not iPods but reward points or frequent flyer miles.

Credit cards, properly used, are marvelous tools. It's a lot more convenient to pay by credit card than to pay by cash; it's a lot easier to write one check to the credit card company per month than a bunch of separate checks to different organizations—you save both time and postage. The credit card companies also have a lot of tools to enable you to organize and audit your expenses. Yet there's a dark side to credit cards, and most people are aware of it. Once you get in over your head and cannot pay the full amount, the interest on the unpaid balance, as well as other penalties, can range from exorbitant to usurious. Shylock and the Mafia might only wish they could do so well.

At the moment, interest rates are relatively low, and my credit is relatively good. I just received an advertisement for a credit card in the mail. Let's look at some of the key provisions. If you receive such solicitations, your provisions will be similar in many respects.[2]

- Annual percentage rate (APR) for purchases—8.99%. I guess they think we won't notice it's really 9%. The prevailing interest rates at the moment are about 3%. The good news is that if you pay the full balance within the grace period (for this card, 25 days after the due date), you won't be hit with any unpaid balance charges. Nonetheless, if you don't pay the full balance, you will be charged interest. How much depends on the method of computing the unpaid balance.

- Balance calculation method. This refers to the way in which the interest you owe is computed. Many cards do this by charging interest on the current balance. For instance, if you owe $500 and pay $400, the fair thing to do would be to charge the exorbitant interest rate on the unpaid balance of $100. Not so with the current balance method; you would pay the exorbitant interest rate on the full $500. The card that I was sent uses the two-cycle average balance method; it computes the average daily balance over two billing periods and you pay interest on that amount.

- Late and overlimit fees—ka-ching! This card charges $19 when the payment is late and the balance is less than $200 and $39 when the balance is more than $200. Ouch! It's $39 if you go over your credit limit during a billing cycle. But here's the real killer: if you ever make a late payment or go over your credit limit, the card has the right to increase the APR up to a maximum value known as the default APR, which for this card is approximately the prime rate + 25%! Of course, the credit card company is not going to do this for the first offense; it wants to keep you as a customer, and chances are if it upped you to the default rate, you'd make your next payment and cancel the card or simply not use it. Do this a couple of times, though, and you could find yourself so far behind the eight ball that digging yourself out could take years.

Last, but very definitely not least, how you treat your credit card will determine to a large extent how lenders will treat you if you want to buy a car or a house.

So what's the lesson? Make damned sure you can pay off the balance on the credit card. Don't make late payments. Don't go over the limit. If you have to make a minimum payment, reduce your use of the credit card as much as possible during the next billing cycle.

My wife, Linda, is a genius with credit cards. We have several different credit cards; she knows what they charge, when they're due, and when we should switch credit cards to take advantage of "sales" that credit cards occasionally have, where they increase the number of bonus points they give to attract new customers or induce old ones to use the card more.

Having multiple credit cards can be advantageous in other ways. Let's look at the example given previously, where you ran up $500 in monthly charges but could pay only $400. If you had five separate credit cards (remember, this is only an example—we have three or four, and I don't know what the nationwide average is) and ran up $100 on each of the five cards, you could pay off four of them and would have to pay interest rate costs on only $100, rather than on $500, and you could choose to make the minimum payment on the specific card that would be most advantageous to you.

Finally, once in the proverbial blue moon, you can actually make money with credit cards! Credit cards often have a provision for making you a cash advance; most of the time they charge you for this. Occasionally (when the moon is blue), they will offer this service temporarily for free. Borrow the maximum amount they allow, buy short-term paper at your local bank with the cash (or stick it in an interest-bearing account, which is pretty much the same thing), and simply pay back the loan during the next billing cycle. For instance, if your card enables you to borrow $1,000 and you can keep it for a month at 3%, that's $2.50. Latte grande: $2.50. Putting one over on the credit card companies: priceless.

How Math Can Help You Buy a House

For most people, the single most important financial decision they will ever make is to buy a house. A house can provide three types of shelter: material, emotional, and financial. Many a young couple will buy a house, live in it as the family grows up, and find that it is paid for when the children move out. As a result, they can live comfortably off pensions and Social Security without having to worry about making payments. The house is a significant financial asset for the estate, enabling the present generation to make life better for the next generation.

As a result, many people push their financial envelope in order to purchase a house, urged on by the multitude of industries that houses support: real estate, construction, and insurance, to name just a few. For some, this decision turns out to be the first major step on the road to financial independence. For others, however, buying a house turns out disastrously—and when too many people make disastrous financial decisions in this area, the entire economy can suffer, as happened recently with the great subprime mortgage fiasco. I'll discuss the impact to the economy in more detail in chapter 10.

The Great Refi Myth

Whenever interest rates decline, you can expect a seductive array of pitches to refinance your loan—any loan. Indeed, as I went to the bank yesterday to deposit a check, all of the vice presidents were wearing pins that said, "Refi now! Ask me how!" Banks want you to refinance, and when interests rates decline, they are able to offer you superficially attractive refinancing packages. Yet beauty, in this case, is skin-deep. Let's see why.

Suppose you are buying a house. Do you want to borrow money at a low interest rate or a high one? Dumb question, right? Of course, you want to borrow at a low interest rate because your future payments are smaller than if you borrowed money at a higher interest rate. So if you borrow money today and interest rates go down significantly tomorrow, wouldn't you be kicking yourself that you didn't wait one more day in order to borrow at a lower interest rate? Of course, you would.

So, how does the fact that interest rates decline a year from now or three years from now make a difference? The answer, of course, is that it doesn't. If interest rates decline at any time from the rate at which you originally arranged financing, you are "stuck" with a series of future payments at an interest rate higher than the one that prevails in the market. How can this possibly be good for you? Of course, it's not.

What the refi industry has done is mathematical sleight-of-hand. When interest rates decline, it is possible to refinance in such a way that your total payments are less than the original plan, and the debt is paid off sooner. The sleight-of-hand comes in convincing you that both of these are good things: less money goes out of your pocket, and you own the title to the house sooner. This isn't technically a scam (which is why I didn't title this section "The Great Refi Scam"), but any attempt to convince people that something good has happened when it hasn't clearly has to be viewed with a jaundiced eye.

Let's look at a typical refi proposition.

Suppose that five years ago, when interest rates were 6%, the bank loaned you $500,000 for the purchase of a house (that was about the median price of a house in Los Angeles in early 2007), payable in 360 monthly installments (the classic 30-year mortgage) of $2,997.75. Five years have gone by, and you have already made 60 payments, totaling nearly $180,000. Most of those payments have gone to pay off interest; only $34,728.03 has gone to reduce the principal. As a result, you owe $465,271.97. The 300 remaining payments represent an outlay that is slightly short of $900,000. Recently, though, interest rates dropped to 4.8%. If you are willing to increase your monthly payment to $3,019.42, you can pay off the loan in 20 years rather than 25, for a total outlay of about $714,660. Through the miracle of refinancing and thanks to the fact that interest rates have declined, you can end up saving almost $175,000. Is this a great country or what?

It may indeed be a good move for you to refinance, but you must understand one very important fact: you are not *saving* $175,000, you are making $175,000 less in payments. It may seem like the same thing, but there's a very important difference, and in order to appreciate what's happening, we need once again to compute present value.

If you do this for your original loan (360 monthly payments of $2,997.75), you're in for an unpleasant shock. You borrowed $500,000 five years ago and have paid the bank nearly $180,000 in those five years. You have 300 payments remaining, and the present value of those payments at the prevailing 4.8% interest rate is $523,170! If you were to somehow win the lottery or have a rich relative die and leave you a lot of money, and if you simply decided to take the course involving the least amount of work and deposit a lump sum in the bank to pay off your remaining payments, you would have to deposit $523,170. That half-million-plus would sit there collecting interest at 4.8%, and every time a payment was due, a check would be issued from the account for $2,997.75. Finally, 300 payments later, the $523,170 would be all gone.

Fortunately, since you probably didn't win the lottery, there is an alternative: refinancing. Recall that if you were to pay off your remaining balance right now, you would have to pay only $465,271.97 because you have paid off some of the principal. There are firms out there that will pay this off for you, and then make you a loan for that amount (more or less, because there's often an early repayment fee) at the prevailing rate of 4.8%. As a result, you can immediately make twenty years' worth of monthly payments of $3,019.42 and pay off the mortgage! It's not magic; it's the way compound interest works. Of course, you could also decide to pay off the remaining balance in twenty-five years, lowering your monthly payments considerably.

At this stage, if you are somewhat cynical, you might ask what's in it for the bank? There's a reason that the loan officers are wearing, "Refi now! Ask me how!" pins. Every time the bank moves money, it makes money. One way of doing this is to charge you an upfront fee for arranging the refinancing; the bank can do this and you can still end up making a lot less in total payments.

It's not a scam, and it may indeed be a good plan for you to refinance. But don't kid yourself that money grows on trees, and that the lowering of interest rates is a bonanza for you, because it's actually the reverse. If interest rates increase after you originally took out the loan, the present value of your remaining payments would be less than the unpaid balance on the loan. If that's the case, you might be able to make out like a bandit, especially if housing prices have gone up (you may remember those wonderful days when that was a foregone conclusion). You could sell your house and pocket the amount that your home has appreciated.

The Great Subprime Mortgage Disaster

The construction industry can't build houses if there are no buyers. When the supply of primo buyers (those with good credit and good credit histories) is exhausted, and there is inventory to be moved, credit will often be extended to those with average credit or worse. Sometimes much worse.

When a lending agency, such as a bank, loans money to an individual whose credit is less than exemplary, there is obviously a greater probability that the borrower will be unable to make the necessary payments. The bank is only interested in the total amount of money that it takes in on the amount loaned out, so in order to compensate for the borrowers who default, it raises the rates to the group at large so that it can take in more interest from the fewer borrowers who can actually make the payments. In doing so, it exacerbates the problem, because an individual who is a marginal borrower at prevailing rates may be pushed over the edge at higher rates.

In the last year and a half, the United States has suffered economically from the effects of accelerated subprime lending. The primo buyers are those who, thanks to their good credit, get the prime lending rates. The others are subprime buyers, who get lending rates above those given to those with the good credit histories.

The desire to own a piece of the American dream, one's own house, is just as strong in a subprime buyer. Yet when the actual costs of mortgage payments needed to purchase a house at subprime lending rates are disclosed to prospective buyers, many will realize that they simply can't make the payments. The banks needed to look for ways to get subprime buyers to ink the deal. As a result, two separate inducements were offered to prospective customers.

The Teaser Rate Trap

Some borrowers were lured into buying houses by being offered "teaser rates." Teaser rates are interest rates well below the current market interest rate; a teaser rate of 1% might be offered for the first two years when actual interest rates are 6%. As a result, the first two years of payments are quite modest; with a teaser rate of 1% on a $500,000 loan, the initial payments are on the order of $1,600 monthly. After the teaser rate period expires, the loan converts into a standard subprime mortgage with considerably higher payments. For many, the teaser rate was barely within reach, so the subprime rates would be out of reach. Even though many of the borrowers were aware of this—and certainly the banks were—all of the parties felt that they would be able to survive. Some borrowers felt this way, but it was based solely on wishful thinking. Others probably felt that their payments would go toward building up equity in the house, so they could either refinance or get out of the situation, based on the equity they had built up. Had they done the math, they would have seen that this was a serious error.

Let's suppose that one has a $500,000 loan to pay off and plans to do so at 6% interest compounded monthly. As we have seen, this can be done in 30 years by making monthly payments of $2,997.75. Yet each month that goes by requires an interest payment of one-half of 1% of $500,000, which is $2,500! So, if you make payments of less than $2,500 monthly, not only will you *never* pay off the loan, the loan balance will increase, which is one reason teaser loans are applicable for only a limited period of time, generally one or two years. As a result, people who were barely able to make payments at the teaser rate found themselves having to make payments that were twice as high, or more, after the teaser rates expired.

And what of the banks? It might be thought that the banks can just foreclose the homes and resell them. When a borrower defaults, however, the bank must face the fact that it received an inadequate return on its loan while the teaser rate was in existence, although the bank will still make a small rate of return. In addition, banks can take advantage of the fact that after some payments have been made on the loan, the loans can be repackaged and sold. As against this, the collapse of the housing market could leave the bank stuck with a $500,000 loan on a house that might be worth only $400,000 now. It's not a pretty picture, whether you look at it from the standpoint of the borrower or of the bank.

The ARM Trap

An ARM, or adjustable rate mortgage, is something like a teaser rate, in that the initial payments the borrower makes are at a rate below the prevailing interest rates. In a fixed-rate mortgage, the monthly payments are the same, regardless of whether the prevailing interest rates go up or down: a borrower is said to "lock in" a fixed rate. In an adjustable rate mortgage, the monthly mortgage payments can fluctuate in accordance with the prevailing rates. This can actually be a good thing if interest rates decline, but if they increase it can be a disaster. As has been observed earlier, subprime borrowers are often stretched to the limit by their payments, and increases in the prevailing interest rates can push them over the precipice.

There's no question that an ARM can be a good idea. In general, ARM rates are initially lower than rates on fixed-rate mortgages. Home buyers who can be assured that their incomes will increase significantly after a short period of time can take advantage of the initially lower rates, knowing that they will be able to afford the higher rates later because of their higher incomes. Those individuals who can see that interest rates will stay the same or head lower will also do better with an ARM— but such individuals would probably do even better if they used such knowledge to clean up using interest-rate futures in the commodities market.

One piece of advice that is sometimes given is that if you plan to live in the house for five or more years, get a fixed-rate mortgage. If you plan to own the house for only a year or two, however, get an ARM, because the initial rates on ARMs are low in comparison to the rates on fixed-rate mortgages, and you'll have it only for the duration of the initial rates. That sounds like good advice, as long as you can follow through on your plan to resell the house. If for any reason you may have to keep possession of the house, you could get clobbered—especially if interest rates start to climb rapidly. It's a gamble: most of the time you win, but when you lose, the loss could have extremely damaging consequences.

Flip and Grow Rich

To "flip" a house is to buy and sell it quickly, making a quick profit. There are two ways this can be done, but only one involves financing. Find a house that merely needs cosmetic work, and do it yourself (or pay for it to be done). Then sell it. Certain improvements can be made to a house that will increase its value significantly more than the cost of the improvement. Cosmetic changes are one such improvement; adding another room or building a swimming pool requires more work but is in the same category.

Most "flippers" tend to rely on the generally true proposition that real estate values increase more rapidly than inflation does.

This is especially true for "attractive" areas: those that are in upscale neighborhoods or locations that are desirable for other reasons, such as being on or near water. Sometimes real estate appreciates very rapidly, and the flipper not only makes money but looks like a genius. It doesn't take much to get into the flipping game: you need the down payment, reasonable credit, and enough money to make payments for a couple of years (most flippers don't hang on for much longer than that), and the results can be very impressive. Ten percent down will often buy a house: if you buy a house for $300,000 and you can get a teaser rate of 2%, payments will be $1,000 a month. Rent the house to someone for a year, sell it when the price goes up to $350,000, and you've made well over 100% on your money in a year. Parlay this for several years, and you'll be on easy street.

The problem occurs when the real estate market stalls out or, even worse, declines. You have to meet the payments; otherwise, you will run into the dreaded F word: foreclosure. Flippers who hit such periods can end up so broke, they have to declare bankruptcy. The good news is that such periods historically last for only a relatively short period of time. So if you're a flipper, you're a favorite to make money, but if you hit a bad period, you're in a lot of trouble.

For most people, though, the venture into housing will prove successful, providing a place to live for a long period and security for the latter portion of their lives. Buy a house intelligently; get a fixed-rate mortgage you can afford, and you have taken a major step toward realization of the American Dream. It is no coincidence that almost seventy percent of Americans own their homes and that America is the richest country on Earth.

Should You Buy a Hybrid?

A lot of other decisions are not as dramatic as buying a house but can be made more intelligently and profitably if you simply do the math. In early 2008, gasoline was rapidly approaching $5 a gallon—at least, in Los Angeles—and many were tempted to buy a hybrid, whose high gas mileage made them one of the hottest-selling cars in the country. Let's take a quick look at whether it makes financial sense.

Suppose that you have a choice between buying a car with a standard engine, which gets 30 miles per gallon, and buying a hybrid, which gets 50 miles per gallon. The hybrid costs $6,000 more than the standard engine. Is it worth it?

One way to analyze the question is to calculate how much gas must cost a gallon to make it economically worthwhile for you to buy the hybrid. To do this, you need to know approximately how long you will own the car (because you have to make up the $6,000 difference, and the longer you own the car, the easier it is to do this) and how many miles you drive annually. Let's assume you plan on owning the car for five years and you drive 12,000 miles per year.

Because you have five years to make up the $6,000 difference, this averages $1,200 a year. If you buy the car with the standard engine, you will use 12,000 miles/30 mpg = 400 gallons of gas annually. If you buy the hybrid, the same computation shows that you will use 12,000 miles/50 mpg = 240 gallons of gas. This means you will save 400 − 240 = 160 gallons of gas annually by buying the hybrid. In order for 160 gallons of gas to be worth $1,200, the price of a gallon of gas must be $1,200/160 = $7.50. I actually did this computation, because I plan on driving my next car about 12,000 miles annually and keeping it for five years, and I settled on the car with the standard engine.

This computation is what mathematicians call a "first-order approximation," because there are other, less important factors that affect the price. Other factors are the cost of charging the battery and the cost of replacing the battery. These costs will vary with the car model, but a hybrid car battery costs several thousand dollars (at the time of this writing).[3] It can last for 100,000 miles or more. If the battery costs $3,000 and it lasts for 100,000 miles, that adds a cost of $.03 per mile, or $1.50 per gallon. Computations such as this only added to my reluctance to buy a hybrid—and I wouldn't even consider it at the current gas price of about $2.25 a gallon. Trendy though hybrids may be, they look like a bad economic bet at the moment. It never hurts to do the math.

A Return Visit to Salina

As we saw in the previous chapter, many crucial decisions can be made simply by doing some elementary arithmetic. The eighth-graders in 1895 Salina would have had no difficulty making a decision about whether to buy a hybrid—only they wouldn't be buying automobiles, they might be buying fertilizer. The fertilizer company charges $600 more for fertilizing a 40-acre farm using enriched fertilizer, which produces 50 bushels of wheat per acre, than for using standard fertilizer, which produces 30 bushels per acre. Is it worth buying the enriched fertilizer?

It's pretty much the same problem as deciding whether to buy a hybrid. Whether it's worth it to buy the enriched fertilizer depends on how much you can get for a bushel of wheat. The enriched fertilizer will generate an extra 20 bushels of wheat per acre, or 800 bushels of wheat for the entire farm. These 800 bushels of wheat must generate at least $600 in revenue to make it worthwhile; this would mean that wheat would have to sell for $600/800 = $.75 per bushel.

Suppose you were a farmer and it was time for the spring plowing, and you had to decide which fertilizer to use. The crop won't come to market until summer. Of course, you'd take a look at the past history of the price of wheat and take your best guess based on that information, but what you would *really* like is a guaranteed price of $.80 a bushel when you deliver your wheat this summer. Welcome to the world of the futures markets, where you, as a farmer, might be able to sell your entire crop of summer wheat at that price, but do so before you have to decide which brand of fertilizer to use. This benefits the consumer as well, because you will then choose to use the enriched fertilizer, putting more wheat on the market. This also brings into existence a new breed of entrepreneurs, the wheat speculators, who will try to profit on price differences in the markets or by guessing which way the price of wheat will move. These individuals are also the ones who, by taking financial risks, enable the farmer to sell his entire crop of summer wheat and make a decision that is in the best interest of the consumer.

Precisely the same arguments could have been made in 2008 for the existence of futures markets, which include the oil market. At that time, oil was about $140 a barrel, and loud outcries were heard that speculators had driven up the price. The same comment might be made of the wheat speculator in the previous paragraph who has actually encouraged the farmer to produce more wheat—which the farmer decided to do because he or she did the math.

9

How Math Can Help You Crunch the Numbers

How did statistics help prevent cholera in
nineteenth-century London?

• • •

Why won't Andre Agassi and Steffi Graf's son be
a tennis prodigy?

• • •

Are you more likely to meet someone over 7 feet tall
or someone more than 100 years old?

can't help succumbing to the temptation to begin this chapter
with the most famous quotation ever delivered on the evaluation of data. Benjamin Disraeli, the youngest person ever
to become prime minister of Great Britain, spoke for not only

himself but others when he said, "There are three kinds of lies: lies, damned lies, and statistics." It's a little surprising to hear that from a Brit, because the first of many triumphs of statistical analysis happened in London, and only a few years before Disraeli became prime minister.

Snow in the Time of Cholera

The word *cholera* does not evoke the same level of fear as *bubonic plague*, probably because cholera is not so easily transmitted, but without proper treatment it is every bit as nasty and fatal. Characteristic of the disease is profuse diarrhea, and death can occur in as few as three hours. Fortunately, proper sanitation procedures will prevent cholera, which is why the disease is virtually unknown in twenty-first-century America.

The same could not be said of nineteenth-century London. Several occurrences of the disease hit London during the summer of 1854, and in late August a particularly vicious outbreak happened in the Soho district. By mid-September, the disease had claimed more than five hundred lives. The scientists of the time believed that "bad air" was responsible for the disease, but John Snow, a London physician, thought otherwise. Snow made a map of the Soho area, inking in those houses where cholera had occurred. With the aid of this map, he was able to show that a public water pump on Broad Street was the likely source of the disease. Although at the time the germ theory of disease had not been proved, Snow was able to convince the local town council to disable the pump by removing its handle. Even though many accounts credit Snow with stopping the epidemic, he himself did not feel that this was necessarily the case. As he wrote,

There is no doubt that the mortality was much diminished, as I said before, by the flight of the population, which commenced soon after the outbreak; but the attacks had so far diminished before the use of the water was stopped, that it is impossible to decide whether the well still contained the cholera poison in an active state, or whether, from some cause, the water had become free from it.[1]

Snow's investigations founded the science of epidemiology. Of equal importance is that they brought into focus the key role that statistical inference can play. As Disraeli wryly noted, though, statistics are prone to abuse; such was true then and perhaps even more so now, with the ability to collect and analyze data so much greater in the twenty-first century than in the nineteenth. The goal of this chapter is to present some of the basics of statistics, with the hope that the reader will then be able to determine in which situations statistics are being correctly used, and in which situations they are Disraeli's third kind of lie.

The Two Goals of Statistics

Broadly speaking, statistics has two goals. The first is to summarize data in some sort of easily digestible format. Most data appear in a mind-numbing blizzard of qualitative and quantitative information, and by using various statistical devices it is possible to convey much of that information in an easy-to-understand way.

Consider the humble pie chart. If you had the complete set of information on personal incomes in the United States, you'd have hundreds of millions of pieces of information. By "binning" the data into well-defined sectors, however, we can construct a pie chart that enables us to see at a glance the approximate income distribution in the United States: the fraction of the population that is poor, lower middle class, upper middle class, rich, and fabulously wealthy. Of course, we need numerical ranges to define the "bins," but once this is done, the pie chart tells most people at a glance all they need to know—or at least what the chart's creator would like them to know through his or her selective choice of bins.

The other goal of statistics is to use sampling procedures to evaluate validity. Consider, for instance, one landmark statistical study that greatly influenced behavior: the connection between tobacco and lung diseases, especially cancer. Ideally, one would want to acquire data for every individual in the United States, find out how long and how intensively that person had smoked, and determine what his or her medical history was. That's simply not feasible. Statistics can come up with a pretty good picture simply by taking a sample of smokers and nonsmokers and finding out whether they have lung problems. Admittedly, the statistical study that enabled the surgeon general to post a warning on cigarette packages was considerably more extensive, but often a simple (and, just as important, inexpensive) statistical study will supply enough information to convince a researcher that there is an important idea worth further exploration. Such a study can also provide evidence that what the researcher thought was an important idea was precisely the opposite, thus (in theory) preventing large sums of money from being thrown away.

How large must a study be to provide reasonably convincing evidence? It depends. In 1998, a survey of approximately twenty type Ia supernovas was sufficient to convince the scientific establishment—a group whose fundamental conservatism makes Rush Limbaugh look like a wild-eyed radical—that the expansion of the universe was accelerating. The last decade has seen a rush on the part of theorists to explain this phenomenon. Past theories have been tweaked and new ones promulgated, all because of a relatively small number of data points.

The Three M's

Perhaps the single most important descriptive statistic is a number that describes where the middle of a set of data lies. What's the average annual income of an American worker? What's the average height and weight of a newborn infant? There are three different ways to measure the middle of a set of data, and they all begin with the letter M.

The first and unquestionably the most important is the mean, or mathematical average, obtained by adding up all of the numbers and dividing by how many there are. If the weights of five newborn infants (in pounds) are 8, 8.5, 9, 7.5, and 9, then the mean is $(8 + 8.5 + 9 + 7.5 + 9)/5 = 8.4$.

The second most important number is the median. When we think of a median in a highway, it's the divider—the strip in the middle—and the median here plays roughly the same role: it's the number in the middle. If we were to arrange the weights of the babies in the previous example in increasing order, we would get 7.5, 8, 8.5, 9, 9. Therefore, 8.5 is the median, the number in the middle, because there are two numbers less than or equal to it (the 7.5 and the 8) and two numbers greater than or equal to it (the two 9s).

Well, it's the number in the middle as long as there is an odd number of data points, such as in the previous example. If there is an even number of data points, it's the average of the two middle numbers. If the previous data are augmented by an additional 8 and arranged in increasing order, we would obtain 7.5, 8, 8, 8.5, 9, 9. The median is then the average of the two middle numbers: the second 8 and the 8.5. So the median in this instance is 8.25.

The reason that the median is much less useful than the mean is that it's a lot harder mathematically to come up with formulas when the quantity you are computing uses different methods of computation, depending on whether you have an odd or an even number of data points.

The last of the three central measures, the mode, is the number or numbers that occur most frequently. In the example of the weights of five infants, the mode is 9; in the example of the weights of six infants, the mode consists of the two numbers 8 and 9. The mode is the worst of the three measures for a variety of reasons. It may not be a single number, and even if it is a single number, it may not be in the middle of the data values. In the example with five weights, the mode was 9, and that's clearly not the middle value in any meaningful sense. If, however, a grocery store wants to make sure it doesn't waste shelf space on displaying the wrong type of pickle, it doesn't compute the mean or the median weight of the pickles sold, but the most frequent (mode) type of pickle purchased. The mode is very useful when the data, such as types of pickles, can't be placed on a numerical scale.

We've barely gotten our feet wet, statistically speaking, and already we can point out one of the reasons for Disraeli's discomfiture. One can get a significantly different picture depending on whether one chooses to use the mean or the median. There was a period a few years ago when oil prices were high and the country with the highest *mean* annual income in the world was one of the small Middle Eastern oil sheikdoms. I'm guessing that said sheikdom probably did not have a *median* annual income anywhere near the figure quoted for *mean* annual income—at least, not if my mental picture of the sheikdom is correct: a few fabulously wealthy oil barons frolicking in the palaces, while the vast majority of underpaid workers sweated to wring the oil out of the ground. At any rate, one can imagine a hypothetical country with one sheik, who has an annual income of $100,000,000 per year, and 99 workers, each with an annual income of $10,000 per year. The mean annual income is a little more than $1,000,000 per year, but the median is $10,000.

There's a wonderful book by Darrell Huff and Irving Geis titled *How to Lie with Statistics*, which goes into matters such as this in much greater depth. Nonetheless, this simple example illustrates that you have to be very careful when confronted with statistical data, to be sure you know what it really represents—and with statistical conclusions, which we'll examine in more detail later in the chapter.

Regression to the Mean

If you're a golfer, you undoubtedly remember the day you shot the best round of your life. You kept the ball in the fairway, avoided the traps and the water hazards, and sank a few putts as well. Maybe you thought that this was your breakthrough, but almost certainly the next day that you went out, you were the same golfer you usually are, spraying the ball all over the place and missing short putts. You've just experienced a phenomenon known as *regression to the mean*. Your usual golf scores are your mean, and unusual scores—both good and bad—are often followed up by average performances. Your unusual scores return, or regress, to the mean that represents the golfer whom you actually are on a day-in and day-out basis.

I have a high ratio of enjoyment to ability in several activities, notably piano and tennis. I also enjoy hearing a great pianist and seeing great tennis players, and two of the greatest tennis players in my lifetime, Andre Agassi and Steffi Graf, got hitched a few years ago. As frequently happens in such cases, they had a child, and speculation ran rampant that this child, considering his parents, could have the makings of one of the greatest tennis players in history.

I can safely say that it's not going to happen. In general, the offspring of two people, both of whom possess an unusual characteristic such as extreme talent or extreme intelligence, will not possess that talent to an unusual degree. This seems to fly in the face of genetics, which discusses how traits are passed on to offspring, but there is a force even more powerful at work here: regression to the mean. Later measurements of data that were originally far from the mean, such as your best round of golf, will tend to be closer to the mean. Just take a look at the batting averages at the end of 2009 for the top ten batters from 2008; almost certainly they will be generally lower, "regressing," or getting closer, to the mean.

This phenomenon was first extended to the field of genetics by Sir Francis Galton in the nineteenth century, who analyzed it in his article "Regression towards Mediocrity in Hereditary Stature."[2] Galton discussed the fact that children of two tall parents tend to be shorter than their parents; similarly, the children of two short parents tend to be taller than theirs.

Regression to the mean has serious consequences for the design of statistical experiments. I'm getting on in years, and cancer of various types is a definite concern. As a result, I tend to notice articles in the paper on the efficacy of various cancer treatments, because if there is a bullet out there with my name on it, I want to know the best options available. I'm especially wary of how the experiments demonstrating the efficacy of these treatments are designed, because regression to the mean can account for a significant portion of a treatment's success if the experiment is poorly structured.

To illustrate, suppose that a new drug for pancreatic cancer is being tested. One thousand people with pancreatic cancer are screened, and the severity of the disease is measured. The bottom 10 percent, being in the most desperate need of improvement, are given this new drug. Regression to the mean dictates that subsequent measurement will almost certainly reveal that the overall severity of the disease has lessened (relative to the average patient in the study) in those being given the new drug, even if the drug consists of extract of bacon cheeseburgers. That's simply because extreme measurements are statistically unlikely; the average scores of the lowest five golfers in the first round of a tournament are very likely to increase in the second round. Of course, a correctly designed experiment will use two groups of subjects. Each group will consist of randomly selected subjects with pancreatic cancer, but one group will be given the new drug and another will be given either the current standard drug or a placebo. Ideally, the study should be *double-blind*; the subjects should not know whether they are receiving the new drug or the standard one, and the doctors administering the drugs should not know which drug each patient is receiving.

How Statistics Have Made Me Immortal

We all know that physical immortality will probably never be achieved by humans, so we have only two viable routes to immortality: through our descendants or through the accomplishment of some memorable achievement. Mathematics and science generally give credit where credit is due; witness the fact that the square of the hypotenuse in a right triangle is equal to the sum of the squares of the two sides will forevermore be known as Pythagoras's theorem.

I have never done any research in statistics, although I do have a pedagogical note on statistics in a small journal. Yet due to an incredibly unusual combination of events, what has been called the most famous result in statistics in the last fifty years shares my name—despite the fact that I had absolutely nothing to do with it. Moreover, I am probably the only mathematician or scientist in the history of mathematics and science ever to have such a result named in this fashion, although I can't be completely sure of this.

One of the staples of entertainment is the "twist" ending: we are led to believe that the playboy son of the rich woman murdered his mother, when the butler actually did it. Or vice versa. The equivalent in mathematics or science is the counterintuitive result, when our intuition about the commonplace leads us astray. We've already seen one instance of this earlier in the book, with the "should you switch doors" example at the game show.

I've been an avid subscriber to *Scientific American* for almost forty years. Every so often, *Scientific American* has an article on mathematics, and in 1977 it had a very interesting article concerning both regression to the mean and future statistical prediction.

The article described an attempt to predict the end-of-season batting averages of eighteen baseball players by using their averages after forty-five at-bats, a common way to predict future averages. The obvious thing to do is look at a player's batting average after forty-five at-bats and predict that he will have the same batting average at the end of the season. The article described the fact that there was a better way to do this by using the tendency of averages to regress to the mean. By analogy with the average scores of the low golfers described earlier, the best batters after forty-five at-bats are liable to end up with lower averages at the end of the season. Baseball fans see this; early in the season there are a few .400 hitters, but the last one to surmount that hurdle by the end of the season was Ted Williams in 1941. The theoretical underpinning for this work was developed by Charles Stein of Princeton University, one of the great statisticians of the twentieth century.

Stein is a fairly common name, both by itself and as a suffix. According to the latest rankings, it's the 720th most frequent name in the United States. We even have a limerick written about us.

> There's a notable family named Stein,
> There's Gert, and there's Ep, and there's Ein,
> Gert's poems are punk,
> Ep's statues are junk,
> And nobody understands Ein.[3]

Even so, we have a way of distinguishing one Stein from another, via first names and middle names if necessary. Mathematics and science do not do this, however. Pythagoras's theorem isn't known as Fred Pythagoras's theorem, only partly because his name was Pythagoras and not Fred Pythagoras. It's Einstein's theory of relativity, not Albert Einstein's theory of relativity; all theorems in mathematics and all discoveries in science are denoted by the last name of the person to whom they are attributed. With one exception—sort of.

That exception is the James Stein theorem, which was the centerpiece of the *Scientific American* article on baseball averages. Okay, it's really known as the James-Stein theorem (because it's a result that was formulated by two mathematicians), but when you say it out loud, no one hears the hyphen. During the course of many years, I have been asked for countless reprints of the article by mathematicians who should know better; after all, there are no other theorems (to my knowledge) that bear the first and last names of the discoverer. Yet I do have a genuine claim to immortality regarding this theorem, rather than the spurious one I just described.

The article in which the James-Stein theorem first appears was printed in 1961 and was authored by W. James and C. Stein.[4] Of course, everyone knew who C. Stein was: Charles Stein, at Princeton. No one knew who W. James was; after this one meteoric flash across the statistical firmament, he had apparently sunk from sight. As you will see, I helped unearth him—although, as you will also see, it wasn't that hard to do.

The James-Stein theorem was the centerpiece of a talk given at a meeting of the American Statistical Association in Los Angeles by Carl Morris, one of the authors of the *Scientific American* article. Here's Morris's account of discovering the identity of James:[5]

Lights were dimmed as I introduced my topic and identified Stein and his work. Then, I offered—ruefully—that statisticians didn't know who James was. A middle-aged man at a rear table called out, "I do!" I could see him only dimly, but I still felt the chill and the premonition that surged through me during the eerie pause that preceded my asking, "Who?"

"I am."

For the next few moments, we conversed one-to-one across the room. Distracted by his appearance, I occasionally would mutter—even during the talk—my amazement that he had appeared. The statistics world finally knew his name. Willard D. James was on the California State University, Long Beach (CSU-LB) mathematics faculty at the time. As a mathematician whose statistics research had been limited to one summer for Stein, he had not kept track of the paper or its full impact. He told the audience he was embarrassed that the estimator Stein discovered was called the James-Stein estimator, and he asked that the "James" be removed to give Stein proper credit. Here are some highlights, mostly learned from our longer, private conversation later that night.

Only a remarkable coincidence brought my ASA talk to James' attention. The CSU-LB mathematics faculty included James Stein (who is still there) [that's me!]. A colleague who spotted the ASA talk announcement with "James-Stein" in the title asked James Stein if that was his work. [Even though he should have known better, as I mentioned, I'm grateful that he did.] James Stein said no, but it was that of Willard James, who was down the hall. So Willard James learned of my ASA talk. And he came.

The Bell-Shaped Curve

The bell-shaped curve is the iconic picture of statistics. Almost everybody knows what it looks like: a rounded symmetrical mountain, the bulk of which lies in the middle. Almost everyone also has a fairly good idea of what it represents. Many traits, such as height, fit the pattern of a bell-shaped curve. The vast majority of people are of average height, maybe a little more, maybe a little less—and only a very few people are either exceptionally short or so tall that they appear to be destined for a career in pro basketball.

The bell-shaped curve, which mathematicians refer to as a normal distribution, derives its value from two sources. The first is that if you have a set of data that fits the bell-shaped curve, you need to know only two quantities in order to be able to answer any statistical question about the set of data: the mean, which is a measure of the middle score of the distribution, and the standard deviation, which is a measure of how tightly packed the data are around the mean.

The formula for calculating the standard deviation is a little complicated, but the smaller the standard deviation, the closer the bulk of the data is to the mean. Switzerland and California have mean incomes that are very close to each other, but incomes are much more spread out in California. There are some extremely rich people (a certain governor comes to mind, as well other celebrities and Silicon Valley billionaires) and a lot of fairly poor people (California has many recent immigrants). As a result, incomes in California have a higher standard deviation than in Switzerland because there are greater percentages of very wealthy and very poor Californians than there are very wealthy and very poor Swiss.

Let's go back to the fact that once you know the mean and the standard deviation of a normal distribution, you can answer any question about it. When you stop to think of it, this is pretty incredible. If I were to tell you that the average score of students on a math test was 77 and then ask you what percentage of students had scores over 85, you would be fully entitled to regard this as a ludicrous question, because you simply don't have enough information to answer it. There are innumerable ways that students could average 77 on a test; in some of them, no one scores over 85, and in others, considerably more than half the class scores over 85. Yet if I were to tell you that (1) the scores on the math test fit a bell-shaped curve, (2) the mean of the scores was 77, and (3) the standard deviation of the scores was 4, everyone who has taken a basic course in statistics knows (or can use a table and quickly find out) that only about 2 percent of the students had scores higher than 85. This specificity, the ability to answer all questions about the distribution from knowing only the mean and the standard deviation, makes the bell-shaped curve valuable because it means you don't have to wade through endless reams of data to answer questions about how the data are distributed; one table will suffice.[6]

The mean and the standard deviation can be calculated for any set of data, and I must admit that the pocket calculator has made this far easier than it used to be (spreadsheets make it a slam dunk). The mean and the standard deviation supply a standardized yardstick for measuring data. In the previous example, a score of 85 is 8 points above the mean of 77. Because a standard deviation for that distribution was 4 points, 8 points is 2 standard deviations, and the score of 85 can be expressed as being 2 standard deviations above the mean.

This standardized yardstick makes it possible to compare data from different environments. Want to know whether you're more likely to encounter someone who is 7 feet tall or someone who is more than one hundred years old? Thanks to the fact that both age and height are normally distributed, you need only find out which number (height or age) is a greater number of standard deviations above the mean.

What makes this standardized yardstick so useful is that it can be used to determine what percentage of the population falls into a given range. We cannot tell what percentage of the population is between six feet and six feet two inches tall simply from these two measurements, but if we translate those two measurements into numbers of standard deviations above the mean, we can.

While I was going to grad school, I worked as a computer programmer for the Educational Testing Service, the organization that administers the SATs. SAT scores are normally distributed, scaled so that the mean score is 500, and each 100 points is one standard deviation. In a normal distribution, half of the scores lie above the mean, about 16 percent lie more than one standard deviation above the mean, and only about 2 percent lie more than two standard deviations above the mean. So the difference between a score of 600 (one standard deviation above the mean) and a score of 500 is considerable—about 34 percent of the scores on SAT tests lie between 500 and 600. The 100-point difference between 500 and 600 represents the difference between an average student and a student in the top 15 percent. The difference between a score of 700 (two standard deviations above the mean) and 600 is substantially less—only about 14 percent of the scores lie in this range, so the 100-point difference here represents the difference between a very good student and an outstanding student. The difference between a perfect score (800) and 700 is very slim, indeed—only 2 percent of the scores lie in this range, so this 100-point difference distinguishes an outstanding student from a truly exceptional one.

The other factor that makes the bell-shaped curve so important is that there are an amazing number of parameters that are normally distributed. I've already mentioned age, height, and SAT scores—the list goes on and on. There is actually a deep mathematical reason for this that is conveyed by one of the most important theorems in mathematics: the central limit theorem of Carl Friedrich Gauss. The basic idea is fairly simple: even though an original distribution of data may be far from normal, if one takes the distribution of means of samples, and if the sample size is large enough, the means of those samples are normally distributed. Thus, if we were to take any set of measurements whatsoever, such as the sizes of shoes purchased by redheads, the individual shoe sizes might be distributed any which way. For instance, because lots of clowns have red hair and very large shoes, we might expect that there would be a surprisingly high number of big shoe sizes—at least relative to the shoe sizes of blondes. If, however, we were to take a large number of thirty-redhead samples, compute the *average* shoe size of the thirty redheads in each sample, and then draw a graph of these averages, we would get a bell-shaped curve. Many real-world quantities are some type of average of several characteristics, and this provides a partial explanation for the prevalence of normally distributed quantities.

What Does It Take to Convince You?

When football season is approaching, you are looking forward to it, not only because you like to watch football, but because you like betting on it—and a friend of yours has developed a system that looks good on paper. Lots of systems look good on paper, however, because they are derived from examining past history, and you have to decide when, and if, to back up your friend's idea with money.

Unfortunately, the system comes up with only one bet per week—and if you test it for too long, the season will be over. So you decide to give it a trial run for one month, betting the games it suggests for four consecutive weeks. It makes bets against the spread, roughly an even-money bet, and you decide that if this system wins four consecutive bets, your friend has really got something, because one normally wins four consecutive even-money bets only 1 time in 16, a little more than 6 percent of the time.

The question "What does it take to convince you?" is at the heart of more than just the science of statistics. It lies at the core of our judicial system, as anyone who has ever served on a jury knows. In order to vote to convict an individual of a crime, you must be convinced of the defendant's guilt "beyond a reasonable doubt." This is the way it is phrased in the instructions that a judge gives to jurors in West Virginia:

> It is not required that the state prove guilt beyond all possible doubt. The test is one of reasonable doubt. A reasonable doubt is a doubt based upon reason and common sense—the kind of doubt that would make a reasonable person hesitate to act. Proof beyond a reasonable doubt, therefore, must be proof of such a convincing character that a reasonable person would not hesitate to rely and act upon it.[7]

That's pretty much the way the judge phrased it when I was an alternate juror. Coming as I do from a quantitative background, I asked the judge what level of doubt was reasonable. One time in five? One time in twenty? One time in a hundred? The judge would not answer this question—at least, not quantitatively—and said it was left for each citizen to decide that for himself.

Statistics, however, quantify what it takes to convince you. The gold standard in "level of doubt" is 5 percent—roughly the equivalent of your chances of making four consecutive winning bets on an even-money chance. Most statistical experiments, especially ones in the social sciences, regard an outcome that would happen less than 5 percent of the time as evidence that there is some underlying reason other than luck for what happened. In the case of the football betting system, the fact that it won four consecutive games would be regarded as solid statistical evidence that the system was a winning one.

The 5 percent level is the "gold standard" for routine social science, but obviously there are situations when it will take more to convince us. The more serious the consequences of making what statisticians refer to as a *type I error*, which in the case of the football system would be to adopt a losing betting system, the lower the "reject" level should be. You certainly wouldn't get in a plane that had a 5 percent chance of not making a successful landing.[8] This level is also much lower for statistical tests of theories in the physical sciences: you can bet the astronomers wouldn't be running around telling us that the universe is undergoing an acceleration of its expansion if they were only 95 percent certain that such was the case.

Hypothesis Testing

Let's see how the "gold standard" is incorporated into hypothesis testing, one of the most important applications of statistics and one that affects our daily lives. Consider, for instance, the television programs we watch—or, rather, the television programs that are available for us to watch. Even as this is being written, one of my wife's favorite television shows (*Fringe*) is "on the bubble," that is, in danger of being canceled, despite the fact that she and several of her friends are avid watchers. Blame hypothesis testing. TV shows are paid for through advertising, and advertising is a classic example of the use of expected value. An advertiser expects to have a small probability of convincing any one particular viewer to buy his product, but if he gets enough viewers, he'll sell a lot of his product. The advertiser has two numbers to determine: the probability that a random viewer of his advertisement will morph into a buyer, and the number of viewers. He can determine the probability through experience or small-scale testing, but he needs statistics to determine whether he can get enough viewers.

Let's try a sample computation. A one-minute commercial costs $15,000; 1 viewer in 30 might buy this company's product; and its profit from each product sold is $5. The number of viewers watching at the time a particular show is broadcast in this particular market is 1,000,000. In order to break even (and obviously this company wants to do more than that), it must sell 3,000 units of its product to pay for the cost of the commercial, and if the fraction of viewers who buy is 1/30, this means that the company needs at least 90,000 people to watch this program, 9% of the viewing audience. Yet the company doesn't simply want to break even, it wants to make a profit, so it decides that it wants the program to attract at least 12% of the viewing audience. Statistics can answer this question only with probabilities. Statisticians set up a straw man they call the *null hypothesis*: less than 12% of the overall viewing audience watched the show. They then ask the question: did we get lucky?

Imagine that we were to put a large number of balls in a jar, 12% of which are red and the rest are white. We then draw out samples of 500, record the number of red balls in the sample, and do this a gazillion times. We next look for the *critical value*, the minimum number of red balls required so that only 5% of the time does the sample contain that number of balls, adhering to the "gold standard" level of acceptance described previously. I've done the computation: 5% of the time 72 balls or more of the 500 are red.[9] The company would have to be tremendously unlucky not to make a profit. This would happen only if less than 12% of the audience watches the program *and* a random sample of 500 viewers happens to contain at least 72 people who watched the program. This combination happens less than 5% of the time.

Incidentally, this is an area in which computers have proved to be tremendously valuable in many diverse areas. This is a relatively simple example with straightforward mathematics, but many procedures are so complicated that it is difficult or impossible to find the correct theoretical value of a certain probability. Computers can be used to simulate the process to find the probability through random trials. A random number generator is used to do this; in the previous example, the computer would generate a random number between 1 and 100; if the number were 1 through 12, it would record the result of a trial as a red ball; if the number were 13 through 100, it would record it as a white ball. Do this 500 times and record the total number of red balls. Then go through this process 1 million times (which on high-speed computers takes only a few seconds); in roughly 5% of the trials, 72 or more red balls will be obtained.

Now let's go from theory to the statistical trial. The Nielsen folks have 500 families monitored in that particular market. During the key period, 75 of them tune in to the TV show. As discussed earlier, under the assumption that 12% of the viewers watch the show, a random sample of 500 viewers would record that 72 or more viewers watched the show only 5% of the time. This is the sign for which the advertiser is looking, and he decides to buy a commercial on the show. There's no guarantee that by doing so he will make money, but if he routinely makes his decisions this way, the "law of averages" will make him come out a winner. Advertising, like life and sports betting, is also one long season.

It undoubtedly seems strange to many people that determining which TV shows live and which ones die is up to only approximately 5,000 households (the Nielsen families), but it's an economical way to ensure success for the advertisers, and most of the time the shows that stay on the air are the shows that America really wants to see.

What Is the "Margin of Error"?

Right before the presidential election of 2008, a typical poll showed that Barack Obama led John McCain by a margin of 49% to 43%, with 8% of the electorate undecided. Just as many advertisements now end with elaborate legal disclaimers, disclaimers are routinely affixed to polls, which usually say something along the lines of "this poll has a 3-point margin of error."

The margin of error is determined in approximately the same way that hypotheses are tested and generally uses the same "gold standard" of 5%. Obviously, we cannot know the percentage of voters who favor Obama until the election takes place. In addition, many surveys are taken for which we will never know the true percentage; when a survey makes a claim such as "72% of Americans, with a margin of error of 3 points, approve of the government's plan to deal with toxic assets," there will be no nationwide vote on the issue. Nonetheless, there is a true percentage for the population at large. If one were to take random samples of the same size as the survey from this population, the percentage of the sample preferring Obama would fall outside the range of 46% to 52% (49% plus or minus 3%) only 5% of the time.

Why Statistics Get the Wrong Results More Often Than They Should

At its core, statistics is every bit as valid mathematically as geometry, yet geometry never gets wrong results and statistics does. Part of the reason that statistics gets wrong results is its inherent nature: 95% accuracy means 5% inaccuracy. So, 5% of the time a network will cancel a show that according to the criteria it had set up, it really should have kept. A new but dangerous surgical procedure will be held to a higher standard, but even if the standard for acceptance is 99.9%, there will still be a few times that a procedure will be adopted when it shouldn't have been.

Yet the majority of errors that occur when statistics are used occur because the sampling procedures are in some way biased, rather than random. Most of the sources of bias are known a priori. Telephone polls are nonrandom because not everyone has a telephone, and even among telephone owners the polls contact only those with land lines, whereas nowadays many people rely exclusively on cell phones. Internet polls are nonrandom because the respondents are self-selected: they ran into the poll in the first place, and they can agree or refuse to take it. Some polls do not measure what they think they are measuring—or, more important, what they want others to believe that they are measuring—because the questions are poorly worded (deliberately or inadvertently) or are designed to elicit certain answers. "Push polls" obviously fall into this category. Finally, respondents to polls may not reveal their true intentions when responding to poll questions. The "Bradley effect," named after Thomas Bradley, a former mayor of Los Angeles, was discovered when Bradley lost the 1982 California gubernatorial race after being ahead in the polls. It was conjectured that white voters may have stated to pollsters that they intended to vote for Bradley, an African American, in order to avoid being stigmatized as racist, when in actuality they intended to vote for Bradley's opponent.

Finally, there is the Picnic Phenomenon (so-called here for possibly the first time). The weather forecast is highly accurate, especially forecasts that do not go more than a day or so into the future, but we tend to remember erroneous forecasts more because it rained on a day for which fine weather was forecast, and we planned a picnic based on that forecast. Everyone remembers the "Dewey Beats Truman" forecasts and the newspaper headlines that were printed bearing inaccurate results, but nobody remembers that the polls showed Reagan winning by a landslide when indeed he did.

My own feeling about statistics is that the less chance there is for human involvement, the more confidence I have in what statistics says. I have a lot of confidence in statistical results in the natural sciences and engineering, a fair degree of confidence when the results are from the life or social sciences, and some confidence—but not a whole lot—when the results concern matters such as elections or marketing decisions. My wife, however, may feel differently than I—because at the last moment *Fringe* was renewed.

10

How Math Can Fix
the Economy

What is the "Tulip Index"?

• • •

What doesn't the mortgage banking industry understand
about negative numbers?

• • •

What caused the stock market crash of 1929?

Health care needs fixing, but our failure to do so has not
done a fraction as much damage to the health of the world
as our failure to recognize that our economy was in an
even more perilous state. This has happened several times
before. In each instance mathematical warning flags were flut-
tering—as early as four centuries ago.

Hey, They're Just Tulips

The first recorded financial bubble occurred during the winter of 1636–1637, in the United Provinces, a country now known to us as the Netherlands (Holland). Tulips had been introduced into the Netherlands during the sixteenth century from the Ottoman Empire and soon became luxury items and status symbols. Tulips have a very short blooming season, in April and May, although the bulbs from which they grow can be replanted until September. This gave rise to two markets: the spot market, during which one could purchase tulips, and a variant of what we would call the futures market, during which growers signed contracts to deliver tulips during the growing season.

All was well with the tulip market until the flowers began to appeal to the French, who increased the demand for tulips. Just as happens nowadays in the various futures markets, speculators entered the picture. The prices of rare bulbs rose to spectacular heights. During the year 1636, a skilled worker might expect to earn in the vicinity of 150 florins a year; at the same time, a single bulb of the Viceroy tulip sold for 2,500 florins.

Prices rise with rarity; it is not difficult for me to accept the fact that a vintage automobile can command fifteen times my yearly salary and that a Monet or another genuinely unique and coveted work of art might fetch ten times what I would make during a lifetime. The financial health of a society is not threatened by this, because there simply aren't that many vintage automobiles or Monets. If, however, there are too many vintage automobiles or Monets commanding these exorbitant prices, there are a number of potential dangers. Some of these dangers moved from the realm of the potential to the actual in February 1637, when the tulip market collapsed.

By this time, tulip trading took place on many of the exchanges located in Dutch cities. Charles Mackay, the author of the classic 1841 treatise *Extraordinary Popular Delusions and the Madness of Crowds*, stated that "the population, even to its lowest dregs, embarked in the tulip trade."[1] Even though some of Mackay's ideas have been debunked by recent economists, it seems clear that tulip trading was the dot-com bubble of 1637 Holland. Possibly at some stage, the lowest dregs looked at one another and said, "What are we doing? Hey, they're just tulips." At any rate, the prices of tulips crashed—and with them went the fortunes of many who had gotten in at the top. The government eventually had to step in, guaranteeing that people who had bought out at the top could get out of the contracts by payment of 10 percent of the contract.

It should be noted that Mackay's view, virtually unchallenged for almost 150 years, has recently come under scholarly scrutiny. Records can be found that seem to indicate that tulip trading was primarily the bailiwick of a relatively small segment of Dutch society, and that the effects of the collapse of the tulip market were limited. Nonetheless, it certainly *could* have happened the way Mackay described, because it *has* happened twice in the last decade, with the collapse of the NASDAQ in 1999 and Alan Greenspan's "once-in-a-century credit tsunami" that inundated the housing market and the banks in 2008. There are definitely common elements to all three of these market collapses, and they relate to arithmetic.

Intrinsic Value

Tulips are certainly attractive, but you can't live in them, wear them, or eat them—despite the fact that once a Dutch sailor, mistaking a rare tulip bulb for an onion, attempted to eat it. The tulip merchant and his family chased down the sailor, berating him for eating up merchandise that, if traded for food, would have fed a ship's crew for a year. The sailor was put in jail.

Value is an elusive concept. Societies have valued tulip bulbs and dot-com stocks all out of proportion to their intrinsic value: what they would actually be worth most of the time. As was pointed out earlier, the problem arises not when the prices of rare items climb disproportionately to their intrinsic value, but when the prices of common objects, such as tulip bulbs or dot-com stocks, do.

Which brings us to 2008.

The Once-in-a-Century Credit Tsunami

One would not think that houses are in the same category as tulip bulbs or dot-com stocks. Houses have a tremendous amount of intrinsic value. After all, whether a house is a small bungalow or a McMansion, we can live in the house. The cause of the tsunami was obviously substantially different from the woes that affected the Netherlands as a result of the tulip market collapse. There were common aspects, though, and they do relate to arithmetic.

I am not the first teacher to notice that my students seem to have very little in the way of a genuine feel for numbers. John Allen Poulos's fine book *Innumeracy* may have been the first literary effort to really do a good job of documenting how pervasive this phenomenon is, but Poulos didn't actually say anything that math teachers haven't known for some time.[2] Some of this can be traced to the ubiquity of calculators: it's a lot easier simply to punch the buttons on a calculator than to understand what the arithmetic operations really represent. Yet part of arithmetic—a very early and fundamental part—is understanding what *numbers* represent. Illinois's venerable senator Everett Dirksen may have brought the issue to the forefront decades ago when, in discussing financial matters pertinent to the U.S. government, he said, "A billion here, a billion there—pretty soon you're talking about *real* money."[3] Multiply those numbers by ten, as a half-century's worth of inflation will, and you're entering the world of the banking and mortgage industries in the current era.

My degree is in mathematics rather than economics, and I dropped Econ 10 after one week, so I certainly do not consider myself able to comment on the underlying economic fundamentals of the credit tsunami. It seems to me, however, that there were three major players in this drama: the prospective house owners, the mortgage industry, and the U.S. government. At the risk of committing some major gaffe in economic theory (my father was right; my dropping Econ 10 might come back to haunt me), let me take a look at the arithmetical and numerical errors of all three parties.

Prospective house owners who found themselves facing the threat of foreclosure generally fell into three categories. The first and clearly the least culpable were those individuals who analyzed the purchase, felt convinced that they could continue to make the mortgage payments through the duration of the mortgage, and then encountered difficulties they had not foreseen. It frequently requires two incomes to support a family nowadays, and when the economy turned sour and one or both of the wage earners lost their jobs, it became impossible to make the mortgage payments. This undoubtedly happens to some extent even under favorable economic conditions, but it obviously occurs much more frequently during recessions.

The second group consisted of those for whom the purchase of a house was a stretch. My wife and I fell into this category. We live in an apartment, and a few years ago our dream house became available. We could have made the payments—we are both mathematics instructors, who are generally in demand (although not at rock-star salaries, more's the pity) even in poor economic conditions, and we are both tenured. Nonetheless, we would have had almost no margin for error, and had a severe health crisis or something else occurred, we would have been in trouble. So we passed on our dream house. Other couples, facing precisely the same circumstances, undoubtedly took the plunge. Some fraction of those ran into trouble and found their houses foreclosed.

The third group consisted of individuals who should not have even been offered the opportunity to purchase a house. Some of the fault here lies with predatory lenders and inattentive government institutions, and this will be discussed later. Yet part of the blame falls on individuals who knew they were living beyond their means. Perhaps certain individuals figured that it was a "heads I win, tails I'm no worse off than I was before" situation. If home values continued to increase, they would be able to obtain loans based on the additional equity; if not, and they could not make the payments, they at least had the opportunity to enjoy a taste of the American dream for a few years.

The first group suffered from bad luck: one simply can't make every decision on a "worst-case scenario" basis, especially if the worst-case scenario seems like a considerable long shot. People in the second group either didn't crunch the numbers or, if they did, chose to ignore them. Sooner or later, if you're living on the edge, an unfortunate event is bound to occur, and the risk is then disproportionate to the reward. My wife and I simply weren't going to risk our future for the chance to move up a little in class. This isn't the easiest calculation to make, but I would expect that many families in our position had essentially the same choice. We could have survived an economic setback living in our apartment but not in a new house. Too much risk, not enough reward.

The last group consisted mostly of free spirits: the "city mice" who never consider that winter may come, or individuals who saw a loophole in the system that would enable them to live considerably beyond their means, even if only for a brief time. Living beyond one's means is either the result of failure to understand negative numbers or understanding them but not appreciating them—sooner or later, the piper must be paid.

Debt and Negative Numbers

I'm insufficiently knowledgeable about the history of mathematics to know why negative numbers were introduced, but I'd bet it has a lot to do with the concept of profit and loss. When discussing negative numbers in either Math for Liberal Arts Students or Math for Elementary School Teachers, I invariably use the idea of debt to illustrate it. The key concept here is that adding a negative number, such as -5, to the positive number with the same magnitude, 5, results in 0. Thus, $-5 + 5 = 0$. Practically anyone above the age of eight realizes that it takes a payment of $5 to cancel out a debt of $5, resulting in the debtor being "all square" with regard to financial dealings with his creditor. Practically anyone, that is, except the second group of people involved in the credit tsunami: the mortgage banking industry.

For most of the lifetime of the banking industry, a home loan was serious business, for both the borrower and the lender. The borrower had to make a sizable down payment even to be considered for a loan. This demonstrated that the borrower had already amassed a certain amount of financial resources and not only was likely to have the ability to repay the balance of the loan, but was unlikely to default early in the loan period, because then the down payment would be forfeited. The bank's money was at stake, and for a long period of time banks were among the most conservative of institutions.

Curiously enough, the image of banks as conservative overlooks a very important aspect of banks: they never have enough cash on hand to repay their depositors! Cash that sits in a bank vault does not make money for a bank; the bank makes money by making loans. Of course, the bank should be prudent, making sure that the loan is likely to be paid off and that the interest generated yields the bank a profit. This ability to loan money allows banks to be engines of credit, providing liquidity to businesses, which enables new businesses to start and existing businesses to expand. As long as times are good, things go well. But when times are bad and faith in the banks disappears (or the depositors need their money), problems occur and there are runs on banks. This was witnessed during the Depression, when many banks had to shut down. The Federal Deposit Insurance Corporation, which guarantees bank accounts up to a certain level (originally $100,000, now $250,000), simply transfers faith from the banks to the government. Perhaps, because the government can print money (and the banks can't), this faith is justified.

In a later portion of this chapter, we will examine the Great Depression and the role that buying stocks on margin played in it.

Loans are liabilities to the borrower, but the bank's accounting system treats them as assets, regardless of how well the loan is secured. If the loans are risky, however, defaults are more likely to occur, with obvious adverse consequences to the bank's balance sheet. One would think that this would impel the banks toward conservative lending practices, as indeed it did in the past. But in the late 1970s, the picture began to change.

The engine of change was the Community Reinvestment Act (CRA), which became law in 1977. It was intended to alleviate the deterioration of low-income and minority portions of cities by requiring that banks meet the credit needs of the community in a safe and sound way. That meant that banks were to devote a portion of their loans to "affordable housing": code for providing home loans to people who might not have ordinarily met the more stringent requirements for down payments and creditworthiness. This particular road to financial hell was indeed paved with good intentions. Being able to afford a house is a key component of the American dream. The intent of the CRA was not only to help people own a home who might otherwise not have been able to afford it, but to promote the pride in community that comes much more easily to those who actually have a stake in it.

Another major piece of the puzzle fell into place in the early 1990s, when Fannie Mae and Freddie Mac, the government agencies that purchased and securitized mortgages, had to devote a percentage of their lending to affordable housing. This enabled banks to hand off their "affordable housing" loans to these agencies, and it proved to be a factor in enabling more, and riskier, loans to be made. Banks were now similar to brokerages, making money on volume, as standards for making loans deteriorated. Mortgage-backed securities were created in which mortgages were bundled together—some good (in terms of the likelihood of the mortgage holder to be able to pay off the loan) and some bad—and these were freely traded. The housing industry, which had been constructed on a foundation of well-researched one-on-one deals between banks and homeowners, had become a market in which bundled mortgages of often questionable quality were traded between institutions and individual investors.

As they say in the market, when the police raid a house of ill repute, they come for the madam as well as the girls. The once-in-a-century tsunami washed away individual homeowners, banks that had made risky loans, and the federal agencies that had underwritten those risky loans. And it all comes back to arithmetic.

The Fluctuating Fractions

Fractions play an essential role in due diligence. The crucial fraction here is a quantity called the loan-to-value (LTV) ratio, which consists of the amount of the loan divided by the appraised value of the property, expressed as a percentage. If a house is worth $400,000 (worth is determined by either an appraiser or a recent similar transaction between a willing buyer and a willing seller) and the borrower wants to borrow $360,000, the loan-to-value is 90%. Obviously, the higher the LTV ratio, the riskier the loan from the standpoint of the lender. There is always a danger that the borrower may default and the value of the house may decrease. That could stick the lender with a loan for an amount greater than the value of the property for which the loan was made.

In the good old days, when due diligence was in vogue, loans would almost never be made with an LTV ratio of more than 80%, because a 20% down payment would be required on the house. Not only did this help ensure the creditworthiness of the borrower, it served as a cushion against a defaulted loan, because the home value could decline by a substantial percentage without harm to the lender.

In October 2002, President George W. Bush gave what in retrospect was a landmark speech on the subject of minority home ownership.[4] He listed four problems that prevented people from obtaining homes: absence of a down payment, unavailability of affordable housing, difficulty in understanding the process and the forms, and difficulty of financing. In line with his philosophy of compassionate conservatism, government programs were either initiated or ones already in place accelerated to make the down payment, to build affordable housing, and to help with financing. In retrospect, these programs were probably too long on compassion and too short on conservatism. Creditworthiness was no longer an issue, because banks received a large measure of support from the federal government. The risk of making a housing loan had, in large measure, been transferred to the federal government—and the more loans a bank made, the more money it made. It all came tumbling down in the late summer and early fall of 2008, when it was realized that Freddie Mac and Fannie Mae, the agencies guaranteeing the loans, were insolvent, and the organizations that had invested heavily in mortgage-backed securities were stuck with "toxic assets," loans that could not be repaid because they were secured by houses whose value had sunk beneath the amount of money loaned on that asset. It will take years, if not decades, to clean up the mess.

A horrifying, yet fascinating, coincidence occurred at the end of 2008, with the juxtaposition of the housing market disaster and the exposure of the multibillion-dollar Ponzi scheme perpetrated by Bernard Madoff. Madoff's scheme was brilliant in its simplicity: he simply paid off early investors at a significantly higher rate of return that was more stable than what was available in the market, but one that was not so high as to raise red flags. Sooner or later, all Ponzi schemes succumb to the necessity of finding ever more investors; at the end, his agents were desperately scouring China to obtain more capital. Similarly, the housing market continued to climb as long as more and more people were found to buy houses, but after the quality buyers had been exhausted, subprime buyers had to be found. When the quality and the quantity of potential homeowners decreased and loans went into default, the house of cards collapsed.

The Math Lesson

The lesson here is not especially deep. I've written elsewhere that numbers and numerical information are meaningful to a segment of the population only as labels: one wears size-8 shoes or lives at 123 Elm Street. The meaning of negative numbers as debt and fractions as LTV ratios should certainly have been known to the mortgage industry, the banking industry, and the government—but they weren't. The disregard of negative numbers and fractions led to the financial meltdown of 2008.

The Crash of 1929

Historians generally agree that the great stock market crash of 1929 was due to a variety of causes, but one exacerbating factor was the ability to buy stocks on very thin margin. Most investors nowadays buy mutual funds for their 401ks, and margin buying doesn't enter the picture. Back in 1929, however, investors did not buy mutual funds, they bought stocks—and they bought them by making a down payment.

To buy a stock on 10 percent margin, one needed only 10 percent of the price of the stock. If a stock was selling for

$50 per share, one could buy that stock with an investment of $5—but one assumed the total risk of the stock's movement above or below $50. If it went up to $55 (and you sold it), you made $5: a 100 percent return on an investment of $5. If it went down to $45, however, your entire investment was wiped out, and if it went below that, you were in debt beyond the value of your investment. Buying on margin was a very risky business, and when the market began to head south, it took the life savings of many small investors with it.

There is an eerily similar parallel between the Crash of 1929 and the tsunami of 2008, which observant readers will probably have noticed. Investors in 1929 leveraged their investments by borrowing money to buy stock and put up the stock as collateral. Shades of 2008, where prospective home buyers borrowed money to buy houses and put up the houses as collateral. As long as stock prices went up in 1929, everything was fine—just as everything was fine in 2008 as long as home prices continued to rise. But in 1929, when stock prices collapsed, the borrowers could not meet the demands for additional capital (the dreaded "margin call") and had to give up the stock to the party that had loaned them the money to buy it. And, of course, in 2008, when the homeowners who had relied on rising prices to enable them to refinance could not do so, their homes were foreclosed. In 1929, those institutions that had loaned money to the speculators were stuck with collateral whose value was less than the amount of the loan—as happened in 2008, when the banks that had loaned money were stuck with houses valued at less than the loan amount. The more things change, the more they remain the same.

Arithmetic and Debt

Negative numbers and fractions are of little interest to most people. They're forced to sit through classes where they learn to manipulate these entities. All too often, the manipulation isn't accompanied by what's really important: attaching these abstract entities to real-world objects. If there are commissions to be had by generating debt, debt will be generated virtually ad infinitum. When a negative number is entered onto the ledger, one of two things will happen: it will either be canceled by positive numbers (when assets increase in value) or it will have to be absorbed by someone if not canceled.

Debt is a two-edged sword. If incurred with due diligence, it enables the real economy to expand by creating additional goods, jobs, and services. If incurred too freely, the resulting financial instruments fail to reflect the underlying values of the real economy, because society cannot produce the assets necessary to pay off the debt—with potentially catastrophic results. LTV ratios are the fractions that stand guard against this happening; as has been seen time and again, we ignore them at our peril.

The Tulip Index

Can mathematics provide us with a warning sign that a meltdown is coming? One possibility is to imagine we are back where this chapter started, in seventeenth-century Holland. When the price of a rare tulip bulb was more than sixteen times the yearly salary of a skilled worker, warning signals should have gone up. When the Tulip Index (the ratio of the price of a rare tulip bulb to the yearly salary of a skilled worker) got abnormally high, it was a sign of trouble. If the price of a Monet exceeds $100 million, it's not a threat to the financial health of society, because there are so few Monets. But when the price of something with little intrinsic value in which a large number of people have a financial interest skyrockets, it's a sign that trouble is brewing.

To construct a modern Tulip Index, I Googled a couple of databases and created the following table. The year 1975 is a baseline, not because of any special significance attached to that year, but because one of the databases didn't have data prior to that year. The value of 1 is arbitrarily assigned to the average annual household income and the yearly close of the S&P 500 for the year 1975; the corresponding numbers for other years in the table are expressed as multiples of the 1975 numbers.

The Tulip Index is a ratio of the S&P yearly close to the annual income, adjusted so that the value is 1 in 1975. Stocks

The S&P Tulip Index, 1975–2007							
Year	Annual Income*	S&P	Tulip Index	Year	Annual Income	S&P	Tulip Index
1975	1.00	1.00	1.00	1992	1.16	4.83	4.14
1976	1.02	1.19	1.16	1993	1.21	5.17	4.26
1977	1.04	1.05	1.01	1994	1.24	5.09	4.12
1978	1.07	1.07	0.99	1995	1.26	6.83	5.43
1979	1.08	1.20	1.11	1996	1.28	8.21	6.39
1980	1.05	1.51	1.44	1997	1.33	10.76	8.11
1981	1.03	1.36	1.31	1998	1.36	13.63	9.98
1982	1.04	1.56	1.50	1999	1.41	16.29	11.54
1983	1.04	1.83	1.75	2000	1.42	14.64	10.26
1984	1.08	1.85	1.71	2001	1.41	12.73	9.01
1985	1.11	2.34	2.12	2002	1.38	9.76	7.06
1986	1.15	2.69	2.33	2003	1.38	12.33	8.93
1987	1.17	2.74	2.33	2004	1.37	13.44	9.76
1988	1.19	3.08	2.59	2005	1.39	13.84	9.92
1989	1.22	3.92	3.21	2006	1.42	15.73	11.08
1990	1.19	3.66	3.07	2007	1.40	16.03	11.44
1991	1.17	4.62	3.96				

*Taken from the U.S. Census Bureau, "Historical Income Tables—Households," www.census.gov/hhes/www/income/histinc/h06AR.html.

are the tulips of our particular era; their value is based largely on what other people feel they are worth, and they have little intrinsic value of their own. Stocks used to be issued to supply capital for a business to start, and people used to buy stocks for the dividend income, but even though these aspects still exist nowadays, most of the money "invested" in stock is basically a bet—like the tulips.

In 1996, for example, the average household income was 28 percent more than it was in 1975, and the closing price of the S&P at the end of 1996 was more than eight times what it was at the end of 1975. That year the Tulip Index stood at 6.39, which means that the S&P had appreciated more than six times as rapidly as annual income since 1975.

Notice that in both 1999 and 2007, the Tulip Index was a little more than 11. It's a curious coincidence, and in both instances the S&P dropped precipitously thereafter, although for different reasons: the dot-com bubble in 1999 and the financial collapse in 2008. A skeptic might argue that this is an artifact of the choice of 1975 as a base year. This occurred to both my editor and me, so I went back into the archives to look at the data from earlier years—at which point I discovered something very interesting. Records for the Dow-Jones Industrial Average showed that the average increased by a factor of about 3.5 between 1929 and 1975. The S&P did a little better: between 1950 and 1975, it increased by a factor of about 4.5. Only after 1975 did the market averages, in particular the broad-based S&P, go into overdrive. What happened to trigger this? This marks, approximately, the beginning of the era of extensive involvement in mutual funds and, equally important, the advent of discount brokerages. Lowered commissions made day-trading feasible: when commissions were high, buy-and-hold was the default strategy. Internet trading exacerbated the situation. Stocks did not become tulips until this combination of events occurred.

The Average Home Price Tulip Index, 1975–2007

Year	Annual Income	Home Price*	Tulip Index	Year	Annual Income	Home Price	Tulip Index
1975	1.00	1.00	1.00	1992	1.16	3.38	2.90
1976	1.02	1.13	1.10	1993	1.21	3.47	2.86
1977	1.04	1.27	1.22	1994	1.24	3.63	2.93
1978	1.07	1.47	1.37	1995	1.26	3.73	2.96
1979	1.08	1.69	1.56	1996	1.28	3.91	3.04
1980	1.05	1.79	1.72	1997	1.33	4.14	3.12
1981	1.03	1.95	1.89	1998	1.36	4.27	3.13
1982	1.04	1.97	1.90	1999	1.41	4.59	3.25
1983	1.04	2.11	2.03	2000	1.42	4.86	3.41
1984	1.08	2.29	2.12	2001	1.41	5.00	3.54
1985	1.11	2.37	2.14	2002	1.38	5.37	3.89
1986	1.15	2.63	2.28	2003	1.38	5.78	4.19
1987	1.17	2.99	2.55	2004	1.37	6.44	4.69
1988	1.19	3.25	2.74	2005	1.39	6.97	5.00
1989	1.22	3.49	2.86	2006	1.42	7.18	5.06
1990	1.19	3.52	2.95	2007	1.40	7.36	5.26
1991	1.17	3.46	2.96				

*Taken from the U.S. Census Bureau, "Median and Average Sales Prices of New Homes Sold in United States," www.census.gov/const/uspriceann.pdf.

A Tulip Index for the prices of houses versus annual income is also revealing. Although not as dramatic as the Tulip Index for the S&P, this table contains its own cautionary tale. The price of homes was increasing rapidly relative to average annual household income, at the same time that more and more people, generally those in the lower regions of average household income, were buying houses. Even though houses have a great deal more intrinsic value than either stocks or tulips do, something had to give—and as we know, it did.

Huge pools of capital throughout the world need to be invested somewhere, and it's hard to see any place for them other than the equities markets. Nonetheless, history has a tendency to repeat itself, and the next time the S&P Tulip Index looks high to me (you can be almost certain this will happen), I'm going to put my money in money-market funds. I may miss the next stock market bubble (this strategy would have missed the dot-com run-up in the late 1990s), but I go by numbers. When the Tulip Index is low, however, I'm going to buy stock. History (not only the data in the table) shows that stocks have generally appreciated more than the rise in inflation, especially during the post–World War II era. I think that a judicious strategy of buy-and-hold (and today, you don't have to pick the right stock, you can buy an index fund), coupled with getting out of the market during times of high Tulip Index, will be a winner in the long run. By March 2009, the Tulip Index was down around 5, and that would have been a very good time to buy the S&P, which promptly went up almost 40 percent in six weeks.

I'll leave it to the stock market prognosticators to work out the key Tulip Index numbers: when to buy, when to get out. I'm not a stock market prognosticator. In retrospect, though, there are a few things I wish I had done in my life that I could have, and one was to set aside a little money each month to invest in an index fund and get out when the Tulip Index is high. Had I done that, I wouldn't be worried about whether the State of California will be able to pay me a pension when I retire.

11

Arithmetic for the Next Generation

How can you get your kids interested in math?

• • •

What is the purpose of arithmetic?

• • •

How does Monopoly money make learning division easier?

've spent my life in mathematics education and have seen all of the trends. Here's my favorite summary of what's happened. I first saw it in the 1980s, and every decade or so since then, updates have appeared:

Teaching Math in 1950 (traditional math). A logger sells a truckload of lumber for $100. His cost of production is 4/5 of the price. What is his profit?

Teaching Math in 1960 (traditional math goes into decline). A logger sells a truckload of lumber for $100. His cost of production is 4/5 of the price, or $80. What is his profit?

Teaching Math in 1970 (new math). A logger exchanges a set "L" of lumber for a set "M" of money. The cardinality of set "M" is 100. Each element is worth one dollar. Make 100 dots representing the elements of the set "M." The set "C," the cost of production, contains 20 fewer points than set "M." Represent the set "C" as a subset of set "M" and answer the following question: What is the cardinality of the set "P" of profits?

Teaching Math in 1980 (the rise of the encouragement of self-esteem). A logger sells a truckload of lumber for $100. His cost of production is $80 and his profit is $20. Your assignment: underline the number 20.

Teaching Math in 1990 (outcome-based education). By cutting down beautiful forest trees, the logger makes $20. What do you think of this way of making a living? Topic for class participation after answering the question: How did the forest birds and squirrels feel as the logger cut down the trees? There are no wrong answers.

Teaching Math in 1996 (teaching math in a bull market). By laying off 40% of its loggers, a company improves its stock price from $80 to $100. How much capital gain per share does the CEO make by exercising his stock options at $80? Assume capital gains are no longer taxed, because this encourages investment.

Teaching Math in 2000 (teaching creative math). A logger sells a truckload of lumber for $100. His cost of production is $120. How does Arthur Andersen determine that his profit margin is $60?

These examples are more than a little tongue-in-cheek, but I'm seeing the results of the 1980 and 1990 examples in my math classes (as I mentioned when discussing the young woman who needed a calculator to take 10% of a number), and I'd like to do something to reduce the chances of this happening again.

This chapter is written for your children or the children with whom you may come in contact. The students I'm seeing in my classroom are basically a lost generation as far as arithmetic is concerned, but there's still time to save the future.

Why Bother?

That's the argument of a lot of students—and unfortunately a lot of teachers—nowadays. Why bother learning how to do arithmetic when the calculator does it so much faster and better? I touched on this in the preface, when I described Isaac Asimov's story *The Feeling of Power*, but let me spell it out so that there is no danger of misunderstanding. The greater a person's comfort level with arithmetic—with counting, comparing, adding, subtracting, multiplying, and dividing—the greater will be the chances of his or her success in higher-level math classes. Perhaps even more important, the greater will be this individual's ability to deal with the mathematics necessary for everyday life.

Algebra is now touted as a gateway course. It's certainly true that if you have problems with algebra, you can probably say good-bye to a career in engineering or the physical sciences, and your chances of success in the life or social sciences are diminished as well. If you have *serious* problems with algebra, you're not going to make it out of high school in a large number of states, because a requirement for graduation is passing an exit exam whose mathematics component is largely Algebra I. Yet I have *never* encountered a student who said, "You know,

I was really good at arithmetic, but algebra totally baffles me." It's just not possible. Algebra involves a level of abstraction and symbolic manipulation that is not generally seen in a standard arithmetic course, but familiarity with numbers and how to manipulate them is the bedrock foundation on which algebra rests. Give me a student who's comfortable with arithmetic, and I'll show you a student who will pass the algebra section of the high school exit exam—unless he or she is sidelined by factors that are totally extraneous to the school environment.

Start Them as Soon as Possible

I don't know whether this is still true, but when I was teaching Math for Elementary School Teachers in the 1980s and the 1990s, a survey of first and second graders revealed that their favorite course was math. By the sixth and seventh grades, math had fallen to last place.

Here's my explanation for that: There is an orderliness and a rationality to arithmetic that children can appreciate. They can work problems and know they're right even before the teacher tells them. Psychologists tell us that children actually want the security of having reasonable boundaries set for their behavior, and arithmetic provides intellectual boundaries. It fits naturally with their needs and wants.

By the time children reach sixth and seventh grade, however, they may feel the need to rebel. If arithmetic isn't placed in some sort of useful context, it becomes a boring series of rote manipulations. Who needs that? This is when children ask, "What's the point of learning long division if a calculator can do it better and faster?"

So the first thing to realize is that the sooner you start helping a child with arithmetic, the better.

The Single Most Important Fact about Arithmetic

It's a skill, and like all other skills, from writing to water-skiing, your ability to perform it improves with practice. To develop your ability at water-skiing, however, you need a large body of water, a motorboat, and water skis. To improve your ability at arithmetic, all you need is the desire and some numbers. Fortunately, numbers are all around you, because arithmetic is the language of quantitative relationships.

And there's one place in particular where you can find lots of numbers.

The Second Most Important Fact about Arithmetic

Using arithmetic is how you deal with money. Newsflash: kids are fascinated by money. It buys stuff, and kids like having stuff.

Back in nineteenth-century Kansas, people knew that one of the most important ways that they could prepare children for life was to make them familiar with the mathematics needed for commerce. Go back to the introduction and look at the 1895 exam again; it's all the arithmetic of commerce. Many things have changed since *Little House on the Prairie*, but not that. Another thing that hasn't changed is the fact that day-in and day-out commercial transactions provide an excellent opportunity for increasing one's arithmetic skills. In fact, it's even more true today for several reasons. A typical day contains many more commercial transactions than it did in the nineteenth century and they are more complicated and use larger numbers. I read *Walden* when I was in high school, and to this day I remember that Thoreau's expenses for a year were on the order of $28.

A tankful of gas costs more than that these days. The arithmetical downside of inflation is that it's a little harder to find arithmetic problems involving money in the real world for the really young child, but it's a lot easier to find them—and there are a great variety—for children in the third grade and higher.

Obviously, there is a great deal of difference in what is to be expected of a child as far as arithmetic proficiency is concerned; it varies with the individual child's ability and the state in which the child receives his or her education, because states control education. In the late 1990s, California adopted a framework for mathematics instruction.[1] I played a very small part in this process, and I think the standards are sort of like the Kellogg-Briand Pact for world peace: great as an ideal goal, not so good as far as actual execution goes. Yet it gives a good game plan for mathematics education, and you can consult it to see how your child is doing and what the appropriate grade level is for a particular topic. Remember, arithmetic is a skill, and the development of a skill is enhanced by doing challenging tasks at the upper edge of one's ability, but it is frustrated by doing tasks that are too difficult. You don't begin learning piano with Beethoven's *Moonlight* Sonata.

So much can be done to help your child with math that libraries are devoted to those books, for those of you who are into the old media, and Web sites flourish by the hundreds for parents and children who prefer the new. One of my former students, Karen Davis, has become a one-woman industry in this area. She has written five books and is the designer of the CoolMath.com Web site, which she informed me with justifiable pride is now the three hundredth most popular destination on the Web (probably ranking only behind porn and YouTube). It receives more than 400,000 unique visitors daily! She started CoolMath shortly after she was a student at California State University, Long Beach (where I teach). Karen is a rare combination of artist and nerd (I don't think she would object to that description). I am also proud to say that I have an original

Karen Davis hanging in my office. It is a graphic design of the face of a vintage calculator (circa 1997), except that in the calculator window where the results of the computation are normally displayed, there is a photograph of an astronaut on the moon. At any rate, CoolMath has lots of stuff for students, their parents, and their teachers. Bookmark it. Even better, visit it with your kids.

You're reading this book, however, so let me at least get you started in helping your children improve their skills in arithmetic. You'll learn how the basic operations of arithmetic apply to everyday commerce and how to become more comfortable performing those operations. Then go check out CoolMath.

Total Recall

That's how we total up a bill. By the time children finish first grade, they are expected to know what are now called "addition facts" (you probably know this as the addition table) through sums up to 20, which means that they should know the sum of two one-digit numbers without having to count the sum on their fingers. There are lots of two-item purchases, such as a burger and a soft drink at your neighborhood fast-food emporium, and if you want to include an order of fries, well, the California framework expects a child to be able to add three one-digit numbers. If any of those items is more than a one-digit number (when rounded off), either you live in a much different neighborhood from the one I do or 1970s-style inflation has set in.

Of course, you never see an item in a fast-food restaurant going for $2—it's $1.95 or $2.49, or something like that. By the time children finish second grade, they are expected to know and use decimal amounts of money for simple problems of addition and subtraction. In my opinion, it's never too early to introduce the concept of rounding as "closer to"; for example, when asking a child to estimate the total cost of a burger and a soft drink, you can say, "$1.95 is close to $2, so let's use $2 to figure out the total. We won't be wrong by much." This idea of estimating an answer rather than coming up with the exact figure is extremely important, because much of the time we only want to get an idea of what a total is. As the child gets older and does more involved computations—by the time he is in fourth grade, he should be able to estimate the cost of an entire trip to the supermarket—you can introduce the idea of compensating rounding. If we've estimated several purchases by rounding up, we might want to round down on the next purchase to compensate for the several purchases on which we've rounded up.

If you don't do mental arithmetic to keep track of your bills, the odds are that you've lost a surprising amount of money during the course of the year just from this. Bills are often incorrectly computed, not because the cash registers are defective, but because the price on the shelf isn't the same price that has been entered into the computer. Stores habitually have sales, and even though scanners can be used to enter the store price into the computer that controls the cash registers, this isn't always done. Here's a fact that will not surprise you: nine out of ten mistakes that are made work against the customer and in favor of the store. I once kept track of how much I saved during the course of a year from mistakes in computing the bill. It was several hundred dollars—in the early 1970s.

Okay, I admit that I have a large component of nerd in me; I like to do mental arithmetic. I like to add up the numerical scores that my students get on the individual problems on an exam in order to obtain the total. But what I really like is when I have mentally kept track of what the bill should be—and there is a large discrepancy between what I think it should be and what it is. Yes, sometimes I'm wrong, but I've saved a lot of money over the course of the years on the times that I am right. If you teach your children to estimate and keep track of expenses this way, they will feel that they are performing a valuable service that helps the family.

There are only two laws of addition: the commutative law, which states that two numbers added in either order results in the same total $(3 + 5 = 5 + 3)$; and the associative law, which says that you can group numbers as you please in order to add them. To add $3 + 5 + 7$, it doesn't matter whether you add 3 to 5 first, getting 8, and then add 7 to this (this is represented using parentheses as $(3 + 5) + 7$), or whether you add $5 + 7$ first and then add that total to 3 (which would be represented as $3 + (5 + 7)$).

Put these two laws together, and you can make addition problems a lot easier. On the first day of Math for Elementary School Teachers, I usually ask my students to add $25 + 89 + 75$ without using a calculator and raise their hands when they've finished. I note the first student to raise her hand (most students in Math for Elementary School Teachers are female, which makes the class a pleasure to teach, because they genuinely want to help children succeed at life), and I conjecture that she once had a job that required her to make change. This is often a good guess, because anyone with that experience uses the laws of addition to group $25 + 75$ (totaling 100), and then add 89 to that total. Encourage your children to use processes like this, and point out regrouping opportunities whenever they occur.

Take It Away

The basic model for subtraction is take-away: if one takes away $3 from a $10 bill, how much is left? This is the change you receive after you make a purchase and hand over the money. It's not as easy to teach kids take-away nowadays as in the past because of the prevalence of credit cards. When making small purchases, however, such as a candy bar or a pack of gum, all you have to do is hand over a dollar, so make sure to keep dollar bills handy. Some stores have minimum amounts for which they will accept a credit card, so opportunities will arise for you to teach subtraction problems involving cash.

You can also create more opportunities to give lessons in subtraction and addition when you give children a weekly allowance. Require them to keep track of their purchases and give an accurate accounting at the end of the week before they receive their next allowance. This accomplishes several laudable goals. It gets your children to appreciate the value of money, to learn to do arithmetic in an environment in which they will use arithmetic in later life, and also to do some rudimentary budgeting—if they want to purchase an MP3 player, they have to save a certain amount each week to do so. It also accomplishes a not-so-admirable goal but one that can unquestionably be useful in later life: the ability to juggle the books. I'm not saying this is a praiseworthy activity, but it does enhance one's ability to do arithmetic.

I mentioned earlier that combining addition and subtraction enables one to perform arithmetical tasks much more simply than one might originally anticipate. I would never directly add $2.34 to $1.88; instead, I would think of $1.88 as $2.00 − $.12 and use this to first add $2.34 to $2.00 and then subtract $.12 from the result, obtaining $4.22. This simple trick is a jaw-dropper in almost all of my math classes (including higher-level courses such as calculus)—the vast majority of students will do the problem the way it is written, even if they are not allowed to use a calculator. Possibly they do this so often using a calculator that they think of the problem the way they would do it on a calculator. At any rate, increasing one's comfort level in doing such problems by regrouping techniques will facilitate the ability to do algebra, which also involves grouping and regrouping.

You can instruct your children to think of negative numbers as being like debt; the number −5 would be a debt of $5. The fundamental property of negative numbers—that $5 + -5 = 0$—is easy to grasp in a monetary framework. In other words, possessing $5 enables one to pay it to cancel out a debt of $5. There are other models for visualizing negative numbers, but this is the most practical and is almost certainly the one they used in nineteenth-century Kansas.

Getting a Quarterback

It's good advice for the defense in football, and it's a good way to become comfortable with addition and subtraction of numbers less than 100. When I was young, you could mail a postcard for a penny, but nowadays there's absolutely nothing you can get for a penny unless it's a "buy one at the regular price, get another for a penny" sale. It's getting to the point that you can't do a whole lot with nickels and dimes either; you can't make a call from a pay phone (you can't even find a pay phone), and the parking meters in Santa Monica accept only quarters, as do all of the Laundromats. So, possessing quarters becomes desirable, and there are several ways you can accumulate them.

When making cash purchases, if you enter a store with twenty-four cents in change (a dime, two nickels, and four pennies), you can give the clerk exactly the right amount to ensure that you get at least one quarter back (okay, bad pun in the heading, but I couldn't resist). For instance, if the purchase comes to $1.68 and you give the clerk $2.18, you are entitled to fifty cents in change. Because half-dollars are also almost nonexistent nowadays, you will get two quarters back. Before the advent of electronic cash registers, which compute the change automatically, I occasionally received perplexed looks from the clerk ("What do you want me to do with this?") until he or she realized that I was trying to obtain quarters. I no longer receive those perplexed looks; the clerks simply fish out the amount of change that the cash register indicates—yet another reason for the decline of arithmetic ability in the twenty-first century.

Go Forth and Multiply

The quintessential business transaction is to buy a number of items at the same price. I'd be willing to guess that this is why the operation of multiplication and the multiplication table were invented. Probably some shopkeeper in Babylonia sold eight clay pots at 4 shekels a pot one day and got tired of adding 4 to itself 8 times, because he realized that he'd have to do it again. At any rate, this is where multiplication gets a major workout.

The older children are, the greater the level of complexity they should be able to handle in multiplication problems, as well as in all other operations. The multiplication table forms the basis for simple computations, but anything higher requires children to understand the procedure of multiplication. This brings us to the laws of multiplication and, more important, the algorithm for multiplying numbers with two or more digits.

There are two laws for multiplication that parallel the laws of addition. The order in which one multiplies two numbers doesn't matter: $8 \times 4 = 4 \times 8$. The way three numbers are grouped for multiplication doesn't matter either: $3 \times (5 \times 7) = (3 \times 5) \times 7$. There is a third law, however, the distributive law, which involves both addition and multiplication—and this is the law that children must understand to really appreciate the process of multiplication.

If a hamburger costs \$3 and a milkshake costs \$2, and you buy four of each, you can compute the total in two different ways: either by computing the cost of the hamburgers ($4 \times \$3$) and the cost of the milkshakes ($4 \times \$2$) separately and adding them up, getting $4 \times \$3 + 4 \times \$2 = \$12 + \$8 = \$20$, or by computing the cost of a hamburger and a milkshake ($\$3 + \$2$) and buying 4 of these: $4 \times (\$3 + \$2) = 4 \times \$5 = \$20$. A \$20 bill sure doesn't go as far as it used to. At any rate, the distributive law states that $a \times (b + c) = a \times b + a \times c$.

I think that one initial cause of the decrease in children's enthusiasm for arithmetic is the appearance of the multidigit multiplication algorithm, which takes some time to learn and is the first of the processes that children do that seems to be clothed in mystery. It's worth spending some time with your children to get them to really understand what's going on when you multiply two 2-digit numbers the traditional way—because it really makes complete sense and it's not that hard to learn. Let's look at multiplying 37×43, first by using the various laws of arithmetic, and finally by seeing how the traditional algorithm puts it all together:

$37 \times 43 = (30 + 7) \times (40 + 3)$, no big surprise here

$= 30 \times (40 + 3) + 7 \times (40 + 3)$, distributive law

$= 30 \times 40 + 30 \times 3 + 7 \times 40 + 7 \times 3$, yet again

$= 1200 + 90 + 280 + 21$, regrouping

$= 1591$

Incidentally, I hope you added up the last four numbers in the order $280 + 21$ first, because they group naturally to total 301.

Now let's look at the traditional algorithm.

$$
\begin{array}{r}
43 \\
\times 37 \\
\hline
301 \\
129 \\
\hline
1591
\end{array}
$$

Of course, you've seen this before: it's $7 \times (40 + 3) = 301$. This is just $30 \times (40 + 3)$ with the last 0 not written $= 1,591$.

Once you get past the invisible zero issue (it simply saves time not to write it), there's generally no problem. Incidentally, you can use play money to actually illustrate this process, but it's best to choose small numbers, such as 13×22, so that you don't spend an hour making 43 piles of $37.

I've chosen 43×37 because it's an example of a basic formula in algebra, some of which can be used as multiplicative short-cuts. The formula is $a^2 - b^2 = (a + b) \times (a - b)$, which is used to factor a difference of squares in algebra but can also be used to multiply numbers quickly if you happen to get lucky, as we have in this situation, by having numbers that are expressible as $a + b$ and $a - b$ with easily computed values of a^2 and b^2. In this case, $43 = 40 + 3$ (where $a = 40$ and $b = 3$) and $37 = 40 - 3$, so $43 \times 37 = 40^2 - 3^2 = 1,600 - 9 = 1,591$.

You can use a cute geometric demonstration to show your kids that $a^2 - b^2 = (a + b) \times (a - b)$. Take a square whose sides are of length a, and cut out a smaller square whose sides are of length b from one of the corners; the physical area of this object is $a^2 - b^2$. You now have an L-shaped block consisting of a small rectangle on top of a big rectangle; cut off the small rectangle. You will have two rectangles, one with sides a and $a - b$, and the other with sides b and $a - b$. Move the two rectangles so that the sides of length $a - b$ are next to each other; you will have created a rectangle with sides $(a + b)$ and $(a - b)$, whose area is $(a + b) \times (a - b)$. I'm sure this and zillions of other goodies can be found on CoolMath— and they'll be a *lot* more visual.

One last thing: I think that once children reach the age of ten, they should be able to multiply any two 2-digit numbers *in their heads*; to multiply 73×84, simply think of it as $(70 + 3) \times (80 + 4) = 5,600 + 240 + 280 + 12 = 6,132$. They have the complete sets of lyrics to their favorite songs memorized, so this really isn't too much to ask. If you get good at testing your children this way, a lot of algebraic manipulation will become a breeze for them—it's similar to what they did in arithmetic.

Divide and Conquer

Absolutely nothing you can do to increase your children's proficiency in mathematics is more important than making them comfortable with division. Although "divide and conquer" was a political strategy of the Roman Senate, designed to help perpetuate the Roman Empire, if your children learn how to divide—and, more important, learn what division is used for—they will almost certainly conquer whatever mathematical challenges they may face.

It amazes me how few people really understand the purpose of division. This probably accounts for the fact that the number of people who are comfortable with mathematics is also relatively small. I mentioned earlier that a couple of years ago, I was teaching a course in college algebra in our school's Honors Program. I thought the college algebra course in the Honors Program would be fertile ground to see how comfortable a typical bright high school graduate not pursuing a technical major would be with division. In the quiz that I gave, which focused on percentage problems, I also asked the following question: what is the purpose of division? Out of a class of fifteen, thirteen students gave an almost word-for-word identical answer: division is when you divide the numerator by the denominator.

Would you answer the question "What is the purpose of talking?" with "Talking is when you speak words"? Of course not; you recognize the difference between speaking words, which represents the *mechanics* of talking, and exchanging views, asking questions, expressing feelings—which is the *purpose* of talking. But most people, including my students, don't recognize the difference between the *mechanics* of division and its *purpose*.

The primary purpose of division is to share a number of items as equally as possible. There are two different models for division, but they merely represent two sides of the same coin. One interpretation of the equation $12 \div 4 = 3$ is that when we share twelve cookies among four girls, each girl gets three cookies. In this case, the number of items to be shared and the number of recipients are known; the problem is to determine how many items each recipient gets. The other interpretation of $12 \div 4 = 3$ is that when we have twelve cookies to be shared and we have determined to give each girl four cookies, we can give four cookies to each of three girls, but one girl is left out. In this case, the number of items to be shared and the number of items to be given to each recipient is known, and the problem is to determine the number of recipients.

What if we had only eleven cookies in each of the previous situations? If we try to distribute them equally to four girls, after we've given two to each girl, there are three cookies remaining ($3 = 11 - 4 \times 2$). Either we have to break them into pieces (entering the world of fractions), or we simply content ourselves with saying that each girl gets two cookies (accounting for $4 \times 2 = 8$ cookies), with three cookies remaining; this is written $11 \div 4 = 2$ R 3. Similarly, if we decide to give four cookies to each girl, we'd be able to give four cookies to two girls (again accounting for $2 \times 4 = 8$ cookies), two girls would receive nothing, and there would be three cookies remaining; again written $11 \div 4 = 2$ R 3. Notice that even though the number 2 is the number of cookies given to each girl in the first example and the number of girls who get cookies in the second example, in each case the number 3 is the number of cookies remaining.

Both problems, $12 \div 4 = 3$ and $11 \div 4 = 2$ R 3, represent what was traditionally called "short division." Students generally don't have much trouble with short division, because short division merely entails knowing the multiplication table and being able to look through it. The way to compute $12 \div 4$ is to think of it as the answer to the question "If you were to make groups of 4, how many groups would you make from 12 items?" We know that one interpretation of $3 \times 4 = 12$ is that 3 groups of 4 items each make a total of 12 items. So, solving $12 \div 4$ simply requires knowing the multiples of 4 and realizing that it takes 3 multiples of 4 to make a total of 12.

The Extended Short Division Algorithm

Generally, students start to have real difficulty executing algorithms when they encounter long division. The road to long division will be considerably easier if you first make your children comfortable with the extended short division algorithm, which deals with the problem of dividing a number with several digits by a single-digit number: for example, dividing 368 by 5.

I'll go through this problem twice, the first time by using a money model, exchanging and sharing as necessary. The second time will be by the pencil-and-paper shorthand, and you'll see that it's exactly the same thing. It's a good idea to have some play money available to actually go through this with your child. Make sure that you have a lot of play ten-dollar bills for this example. You can make your own play money by cutting up rectangular strips of paper as needed, if you don't have Monopoly money or something similar on hand.

To divide 367 by 5, start with 367 dollars: 3 hundred-dollar bills, 6 ten-dollar bills, and 7 one-dollar bills. Get five markers to represent people—if you have photos of five different people, you can use those, or else get five checkers or colored pieces of plastic; the markers simply define an area of space into which you will share the money. Now we're ready to start the sharing and exchanging process, which is the purpose of division.

You obviously can't share the 3 hundred-dollar bills equally among the 5 people; there simply aren't enough hundred-dollar bills. Exchange the 3 hundred-dollar bills for 30 ten-dollar bills. You now have 36 ten-dollar bills—the 6 you initially started with and the 30 you just obtained by exchanging—and 7 one-dollar bills. Because $7 \times 5 = 35$ and $8 \times 5 = 40$, you have enough ten-dollar bills to give each person 7, but not enough to give each person 8. So you give each person 7 ten-dollar bills; as we have just seen, this accounts for 35 of your 36 ten-dollar bills, leaving you with 1 ten-dollar bill.

You can't share the 1 ten-dollar bill equally among 5 people, so you exchange it for 10 one-dollar bills. You now have 17 one-dollar bills: the 7 you initially started with and the 10 you just obtained by exchanging. Because $3 \times 5 = 15$ and $4 \times 5 = 20$, you have enough to give each person 3, but not enough to give each person 4. So you give each person 3 one-dollar bills, accounting for 15 of your 17 one-dollar bills, leaving you with 2 one-dollar bills.

You're all done. Each of the five people has 7 ten-dollar bills and 3 one-dollar bills—73 dollars—and you have 2 one-dollar bills remaining. You have just gone through the physical process of showing that $367 \div 5 = 73$ R 2.

Now let's take a look at the pencil-and-paper shorthand; you'll see that every step of it corresponds to something you did previously.

$$5\overline{)367}$$

Just as 367 dollars is 3 hundred-dollar bills, 6 ten-dollar bills, and 7 one-dollar bills, the number 367 is 3 hundreds, 6 tens, and 7 ones. You can't share 3 hundreds equally among 5 recipients, so you exchange the 3 hundreds for 30 tens. Added to the 6 tens with which you started, that's 36 tens. Notice that 36 forms the first two digits of 367. That's the beauty of the base-10 number system that we use: it incorporates the sharing and adding idea in the way the numbers are written. So 367 can be thought of as 3 hundreds, 6 tens, and 7 ones—or 36 tens and 7 ones. Or 367 ones.

Again, because $7 \times 5 = 35$ and $8 \times 5 = 40$, you have enough tens to give each recipient 7, but not enough to give each person 8. When you give each person 7, you use up 35 of the 36 tens, leaving 1 ten. This is written

$$\begin{array}{r} 7 \\ 5\overline{)367} \\ 35 \\ \hline 1 \end{array}$$

You can't share the 1 ten equally among 5 recipients, so you exchange it for 10 ones. Added to the 7 ones you already had gives you 17 ones: $10 + 7 = 17$. The 1 ten that's hanging at the bottom is the same as 10 ones, and when you "bring down" the 7, you are simply condensing the idea that 1 ten plus 7 ones = 10 ones plus 7 ones = 17 ones.

$$\begin{array}{r} 7 \\ 5\overline{)367} \\ 35 \\ \hline 17 \end{array}$$

We're almost done. Because $3 \times 5 = 15$ and $4 \times 5 = 20$, we can give each recipient 3 ones but not 4. Giving each recipient 3 ones uses 15 ones, leaving $17 - 15 = 2$ ones. This is written

$$\begin{array}{r} 73 \\ 5\overline{)367} \\ 35 \\ \hline 17 \\ 15 \\ \hline 2 \end{array}$$

That's the shorthand for $367 \div 5 = 73$ R 2. The physical version can be carried out without any knowledge of the multiplication table. For instance, when you have 36 ten-dollar bills, you can simply distribute them one at a time to the five people; at the end of doing so, each of the five people will have received 7 ten-dollar bills (you can count each person's stack of ten-dollar bills to verify this), and you will have 1 ten-dollar bill remaining. Of course, knowledge of the multiplication table speeds things up considerably, which is why the California Framework requires memorization of the multiplication table to an automatic level.

Long Division

If you are comfortable with short division, you shouldn't have much difficulty with long division, for the idea is exactly the same. The only real differences are that you may find it a bit harder to work out the individual digits, because you can't rely on the memorized multiplication digit to do so. You also need to be comfortable multiplying several-digit numbers by a single digit, but other than that, nothing changes. It's still share and exchange. Let's divide 619,853 by 814; this is about as extensive a division problem as your children will ever encounter. If you can get them past problems like this, you're home free on the procedure front.

$$814\overline{)619853}$$

Because 814 is more than 619, we exchange the 619 thousands for 6,190 hundreds; added to the 8 hundreds in the original number it gives us 6,198 hundreds. If we look at the first digit of 814 and the first two digits of 6,198, we get an idea of what the first digits of the answer would be; this is where familiarity with numbers helps speed up the process. Because 8×8 is 64, then 8×814 would be more than 6,400, which suggests we should take a shot at using 7 as the first digit. We next multiply 7 by 814 and subtract from 6,198.

$$
\begin{array}{r}
7 \\
814\overline{)619853} \\
\underline{5698} \\
500
\end{array}
$$

Each of the 814 people has received 7 hundreds from the original collection of 6,198 hundreds, which accounts for 5,698 hundreds, so there are 500 hundreds left—which we exchange for 5,000 tens. Added to the 5 tens in the original number gives 5,005 tens, which we see when we "bring down" the 5.

$$
\begin{array}{r}
7 \\
814\overline{)619853} \\
5698 \\
\hline
5005
\end{array}
$$

Again, looking at the first digit of 814 and the first two of 5,005 suggests that we try 6 as the next digit. Every so often, this will be the wrong choice; if the divisor had been 844 rather than 814, when we multiply 6 by 844 we will obtain a number larger than 5,005. A student who is comfortable with mental arithmetic will be able to "see" that this will happen and will realize that 6 wouldn't work and will use 5 instead, saving time and resulting in a neater test paper if this problem appears on an exam. This time, however, we're okay with using 6 as the next digit.

$$
\begin{array}{r}
76 \\
814\overline{)619853} \\
5698 \\
\hline
5005 \\
4884 \\
\hline
121
\end{array}
$$

Each of the 814 people has received 60 tens from the original collection of 5,005 tens, which accounts for 4,884 tens, so there are 121 tens left—which we exchange for 1,210 ones. Added to the 3 ones in the original number gives 1,213 tens, which we see when we "bring down" the 3.

$$
\begin{array}{r}
76 \\
814\overline{)619853} \\
\underline{5698} \\
5005 \\
\underline{4884} \\
1213
\end{array}
$$

As you can easily see, we can give only a single one to each of the 814 persons, and this completes the problem.

$$
\begin{array}{r}
761 \\
814\overline{)619853} \\
\underline{5698} \\
5005 \\
\underline{4884} \\
1213 \\
\underline{814} \\
399
\end{array}
$$

So the answer is 761 with a remainder of 399.

Averages

You've seen how important averages are; they occur throughout this book. Having read the chapter on statistics, you know that the mean is the average value of the data under consideration. If you buy 2 pounds of apples for $0.80 a pound and 3 pounds for $0.60 cents a pound, you have purchased 5 pounds of apples for a total of $3.40. The mean value of each pound of apples purchased is $0.68 (= $3.40/5); that's also the average cost of a pound of apples.

We have seen that division represents an equal sharing of items among a number of recipients. In the previous example, the average of $0.68 can be regarded as the result of sharing the

$3.40 total cost among 5 pounds of apples. An average is also a quotient, and a quotient consists of a numerator and a denominator. When we are dealing with real-world quantities, the numerators and the denominators are measured in units. The numerator units in the previous example are dollars, and the denominator units are pounds. To fully understand an average, one must know what is being shared and among what the shared quantity is being distributed. As was pointed out earlier, the units being shared are dollars (the numerator units), and pounds are the denominator units among which the dollars are distributed. The units of measurement for averages are "numerator units per denominator unit"—in this instance, dollars per pound.

In the previous example, the average $0.68 can be regarded as the answer to this question: If five pounds of apples cost a total of $3.40, how many dollars would each pound cost if each pound cost the same amount? Phrased this way, the computation of an average is a division problem: $3.40 ÷ 5 = $0.68.

Yet it is also possible to think of division problems as multiplication problems in reverse; if each pound of apples costs $0.68, 5 pounds would cost $3.40. This is expressed by the following equation: 5 × $0.68 = $3.40.

If you think of $3.40 ÷ 5 as the number to be multiplied by 5 to give $3.40 as a result, this is an alternative way to look at division.

The units used to describe an average are a key part of the information that an average conveys. When you compute an average, a number by itself is meaningless: both the numerator and the denominator units must be specified. To see how important this is, ask yourself whether you would take a job if the salary was simply described as "5." Assuming that the job isn't distasteful or dangerous, you almost certainly would take the job if the salary was 5 dollars per second. You most likely wouldn't take the job if the salary was 5 cents per year.

Summarizing the Past, Predicting the Future

In the chapter on statistics, we noted that the mean is the most useful of the measures of the middle, and it is always used to compute average prices. You might hear of the median value of a house, but never about the median cost of a pound of apples or a gallon of gasoline.

Using past averages as future estimates is common practice. Hypothesis testing and confidence intervals in statistics start with a summary of past data, in the form of either a proportion or a mean, and use this as a basis to describe the value of a parameter for an entire population. Think of the data gathered as a sample that's part of a population consisting of the entire set of data—past, present, and future—and you will see how statistics will take past averages and use them to project future values.

Sometimes there are several different approaches to problems involving averages. For example, suppose Bob needs to sell an average of $20,000 worth of computer equipment monthly in order to qualify for a bonus. During the first eight months of the year, he has averaged $18,000 per month. There are two different ways to compute how much he needs to average in the last four months to qualify for the bonus. The straightforward way is to realize that he must sell a minimum of $12 \times \$20,000 = \$240,000$ for the entire year. Through the first eight months, he has sold $8 \times \$18,000 = \$144,000$. He must therefore sell $\$240,000 - \$144,000 = \$96,000$ in four months, for an average of $\$96,000 \div 4 = \$24,000$ per month.

The following approach is a little more sophisticated but makes for easier calculation. Each of the first 8 months Bob has fallen $2,000 short of his quota, so he is a total of $8 \times \$2,000 = \$16,000$ short. He needs to make that up in four months, an average of $\$16,000 \div 4 = \$4,000$ per month. In other words, he must exceed his previous quota by $4,000 per month for the last four months and must therefore sell $\$20,000 + \$4,000 = \$24,000$ worth of equipment per month.

Other averages that are easily computed are parameters associated with your children's lives. What is the average amount of time it takes them to walk from school? What is the average daily number of hours that your children watch television? The average daily number of hours that they spend on the Internet? It isn't necessary to turn your children into data collectors and analyzers (okay, let's say it, geeks). If you feel that changes need to be made, however, you can say something like, "Let's limit your television watching to an average of two hours a day every week," so that your children can plan for special events that they may want to watch that run more than two hours.

Finally, because averages play so large a role in the news, make a habit of going over the news of the day with your children, paying special attention to averages. This serves a dual purpose: it helps reinforce the arithmetic concepts involved (don't forget to stress what are the numerator units and what are the denominator units whenever you see an average), and it is also a civics lesson, making your children aware of current events. By the time your children are able to work with averages, they can also absorb and understand what is happening in the world. Comprehending how averages reflect what occurs in your city, your state, your country, and the world will boost your children's mathematical proficiency and cultivate their progress toward becoming citizens.

One Final Piece of Advice

Obviously, there is a lot more to improving arithmetic proficiency than I've talked about here, but it's like the policeman said to the man carrying a violin who asked how to get to Carnegie Hall: "Practice."

Live long and calculate. Teach your children to calculate. It's a good way to ensure that you—and they—will prosper.

12

How Math Can Help
Avert Disasters

What caused the *Challenger* space shuttle crash?

• • •

How could we have prevented much of the damage from
Hurricane Katrina?

• • •

How can you determine the possible cost of a disaster?

I t seems that the great lessons of life, whether they are lessons
for the individual or for societies, are always accompanied
by pain. When things are going swimmingly, we bask in the
warm fuzzy glow that surrounds success. Federal commissions
are rarely formed to investigate how things went right; they only
investigate how things went wrong. A lot of pain accompanied

three of the great disasters of the last quarter century, and they all could have been either avoided or greatly minimized had someone only done the math.

January 28, 1986

The morning dawned clear and cold in Florida. The *Challenger* space shuttle launch had originally been scheduled for January 22, but a series of delays had occurred, and the launched had been pushed back to January 28, which was also the date of President Reagan's State of the Union address.

The low temperatures were a source of considerable concern for the engineers at Morton Thiokol, the company responsible for the construction of the shuttle's solid-fuel rocket boosters. A teleconference had been held the night before between Thiokol and NASA personnel, with the Thiokol engineers expressing their concern regarding the resilience of the rubber O-rings used to seal the joints on the boosters. This was a serious issue, because the O-rings were a "criticality-1" component, whose failure would result in the loss of *Challenger* and the astronauts aboard. The engineers, however, were overruled by Thiokol management, who recommended that the launch proceed as scheduled.

Prior to the launch, ice buildup on the service structure standing next to the vehicle was also noted—but as the day proceeded, the ice began to melt. The launch was delayed until late morning, and the final clearance was given at 11:38 a.m. Florida time.

I have always been extremely interested in the space program and had planned to watch the televised launch from my apartment in Los Angeles. I had morning classes to teach, however, and had to leave before *Challenger* was launched. Like much of the nation, I found out soon after the tragedy what had occurred.[1] To this day I find myself unable to watch the video of the launch, even though it happened more than twenty years ago.

September 23, 1998

Options are an extremely important type of contract. An option to buy gives the holder the right, but not the obligation, to buy something at a fixed price on or before a certain date. A motion-picture company, contemplating the production of a picture in which Angelina Jolie would be perfect for the leading role, might purchase on option for her services. Such a contract might be structured as follows: the company pays Angelina Jolie $1,000,000 for the right to sign her to a contract to star in the film, for which she would be paid $15,000,000, with the right to sign her expiring on January 1, 2011. The company may end up not making the film, but the $1,000,000 is insurance that if it does, Angelina Jolie will star in it for $15,000,000. It's a win-win arrangement, for Angelina Jolie gets an extra $1,000,000 (like she needs it!) if the film is made and walks away with the million if it isn't—after January 1, 2011, she is free to make other arrangements.

Stock options (the right to buy a stock, such as Microsoft, at a certain price on or before a certain date) have been around for a considerable period of time, but the contracts were originally relatively complicated agreements that were individually negotiated between a buyer and a seller. Stocks themselves were once traded this way, prior to the existence of stock exchanges. Yet interest in stock options as investment vehicles increased during the 1950s and the 1960s, and exchanges for trading stock options came into being in the early 1970s.

At the same time, two brilliant mathematical economists, Fischer Black and Myron Scholes, derived a mathematical expression for the value of a stock option. I was interested in stock options and read the paper in which they explained this result; it involved constructing and solving a differential equation, using assumptions about neutrality of risk that paralleled the assumptions of conservation of energy that physicists and engineers use when they model physical systems with differential equations. This was brilliant—and illustrative of what makes mathematics such a powerful tool: ideas that could be mathematically expressed in the laws of nature have analogues that can be expressed in the world of finance.

The Black-Scholes model was, for a while, the Holy Grail for floor traders at the options exchanges. They would locate two options that were mispriced relative to each other and to the theoretical value suggested by the Black-Scholes model and would buy the option that was undervalued relative to the theoretical value and sell the overvalued one, waiting for the passage of time to bring the prices back into line and enable them to emerge with a profit. Even if this did not work on an individual trade, if the model was accurate over the long run, the law of averages (in this case, the fact that each individual trade had a positive expected value) would ensure that profits would accrue to the trader.

This work led to a Nobel Prize in economics—and to the founding of the hedge fund Long-Term Capital Management (LTCM), whose board of directors included Scholes, Robert Merton (who also won a Nobel Prize for his work in this area), and John Meriwether, who had been a top bond trader at Salomon Brothers. LTCM was brilliant in its conception; academics were to supply the quantitative models to devise trading strategies, and traders with impeccable track records would do the actual trading. The entry price to LTCM was steep—$10 million—but eighty investors each came up with the ante and LTCM started with $800 million in equity. Its basic strategy was to look for securities that were mispriced relative to one another, much as options floor traders did with options spreads. Because these mispricing differences were small, however, LTCM had to take large positions in order to make a small profit—a strategy that was likened to "picking up nickels in front of a steamroller."

The first two years LTCM made 40 percent, and the third year 27 percent. By this time, LTCM was managing $7 billion. Meriwether, however, returned nearly $3 billion to the investors because there did not seem to be a sufficient number of attractive investment opportunities.

At the beginning of 1998, LTCM had $4 billion equity, which, through the leveraging that exists via options and futures contracts, controlled $100 billion in assets. LTCM had also become a huge player in emerging markets, such as Russia, a move that would ultimately prove fatal. On August 17, Russia devalued the ruble and declared a moratorium on its debt. This had a catastrophic effect on LTCM, because it could not make the readjustment moves that its trading strategy demanded, and by September 22 its equity had shrunk to $600 million while it still controlled a huge portfolio. Demands for additional capital, known as margin calls (which played a key role in triggering the stock market crash of 1929 and the resulting Depression), could not be met. Normally, margin calls are satisfied by liquidating assets, but LTCM's assets could not be liquidated. This led to fears that this could act as a trigger to a systemic meltdown, much as the failure of a single relay at the Adam Beck power station in Ontario, Canada, triggered the 1965 blackout that crippled the northeastern United States. On the afternoon of September 23, the Federal Reserve Bank organized a rescue effort by a consortium of investment banks and LTCM creditors that pumped almost $4 billion into LTCM. Four billion dollars may not sound like much, but the timing was crucial, and the meltdown was avoided.[2]

August 29, 2005

Hurricane Katrina was not the strongest hurricane ever to hit the United States, and it was by no means the deadliest—but it unquestionably did more damage and higher-profile damage than any other storm in history. The picture most of us will retain of Hurricane Katrina is that of a flooded New Orleans, inundated as a result of the failure of fifty-three levees surrounding the city. The city that was home to Mardi Gras was almost instantly transformed into something most Americans associate with disasters in Third World countries. Years later, New Orleans still has not recovered, and complete recovery may take decades.

In the aftermath of the storm, numerous investigations were conducted into the cause of the levee failures.[3] The American Society of Civil Engineers, in a June 2007 report, concluded that the failures were due to system design flaws. The U.S. Army Corps of Engineers, which had designed and constructed the system, objected to this report, claiming that Katrina was so strong it would have overwhelmed the levees. This claim was rebutted by investigators from the National Science Foundation, who pointed to a 1986 study by the Army Corps of Engineers that had mentioned the possibility of precisely the failure mechanism that actually occurred.

The Two Key Questions

When disasters of the magnitude of the *Challenger*, LTCM, and Katrina occur, there are plenty of opportunities for second-guessing. This second-guessing generally takes the form of two key questions. The first and most obvious is "Could anything have been done?" Usually, the answer is affirmative. The launch of the *Challenger* could have been delayed. LTCM did not have to invest in Russian assets. The levees could have been strengthened and the design flaws corrected.

The second key question, "Should anything have been done?" is more difficult to answer. All three of these cases represent worst-case scenarios, but if we made all of our decisions on the basis of anticipating the worst, we'd never get out of bed, because we might slip and hit our heads on the bedpost.

Considering that breakfast waits in the kitchen or the dining room, and we have to work in order to have a bed to sleep in or a breakfast to eat, most of us risk slipping and hitting our heads on the bedpost because, after all, this is an extremely low-probability event. But low-probability events *do* happen, and mathematics has a way of assessing the best course of action when risky events with low probability are part of the scenario. It should come as no surprise that an expected-value analysis could have, and probably should have, enabled the correct course of action to be adopted in each of these situations.

An Expected-Value Analysis of the *Challenger* Disaster

In the aftermath of the *Challenger* disaster, practically every aspect of the enterprise came under scrutiny, and almost everyone came in for some share of the blame. There is a legal doctrine called "last clear chance," under which the defendant in a legal action will not be liable if the plaintiff has the last clear chance to avoid the accident. The last clear chance to avoid the accident was clearly in the hands of the NASA controllers, who gave the final go-ahead.

This go-ahead, however, came as a result of poor information and poor communication. Former secretary of state William Rogers headed the commission investigating the disaster. The commission concluded that "failures in communication resulted in a decision to launch 51-L based on incomplete and sometimes misleading information, a conflict between engineering data and management judgments, and a NASA management structure that permitted internal flight safety problems to bypass key Shuttle managers."[4]

The idea that a failure in communication and information presentation was the pivotal component of the accident seems to have worked its way into many of the analyses that have been given of the *Challenger* disaster. It seems to me, though, that this is really not the critical issue. Of course, communication and information presentation are important, but having all of the relevant information would not, of itself, have prevented the disaster. The key point concerns the processing of the information.

True, it is quite likely that had the information "the Morton Thiokol engineers believe that there is a serious possibility of O-ring failure if launch occurs at temperatures below 53° F" been clearly communicated to the launch controllers, the launch would have been scrubbed—or at least delayed. Yet messages such as this can be drastically modified by changing or eliminating a single word. If the word *serious* is eliminated, the message conveys almost no warning at all, and if that word is changed to *distinct*, the strength of the warning is open to question.

Mathematics has an answer for this: use numbers! Communicate warnings in terms of estimates of probability! Suppose that the Morton Thiokol engineers had communicated the information that they estimated a 10 percent chance of catastrophic failure. Because the *Challenger* crew consisted of seven people, the expected value of the launch would amount to the loss of 7/10 of a human life. It is a lot easier to scrub the launch when you realize the cost in human lives. If you also compute the expected value in terms of money, an estimate of a 10 percent chance of catastrophic failure, which might have been qualitatively translated as "a distinct chance," would almost certainly have prevented the disaster.

An Expected-Value Analysis of the LTCM Disaster

I spent some years as a stock-option trader. I'm nowhere near old enough to remember Black Monday—October 28, 1929, when the Dow lost 13 percent in one day—but I am old enough to remember the second Black Monday, October 19, 1987, when the Dow lost almost 23 percent in a single day. I had my own mini-LTCM catastrophe on that day.

I had devised a strategy, interestingly enough somewhat akin to LTCM's "picking up nickels in front of a steamroller," that had never lost in the six months I had traded it. I had a partner at the time, and after he saw this amazing winning streak, he asked me what could cause it to lose. With almost frightening prescience, I told him that the only way I could see that the strategy would lose would be if we were unable to liquidate our positions—and that hadn't happened to my knowledge in the history of the options exchange (which had started early in the 1970s) and not in the stock market since Black Monday in 1929.

Of course, that's exactly what happened. There simply were no bids—no one wanted to buy what we had to sell, and we ended up losing an amount of money that took me eight years to pay off. I learned a fundamental lesson then: when constructing a strategy, you must consider the possibility that you may not be able to execute it. If such is the case, your losses must be limited in some respect. Although I have traded stock options with moderate success since then, I have never even considered strategies that have small probabilities of astronomical losses. I don't know whether this comes from a subconscious expected-value analysis or the lesson "once bitten, twice shy," but such strategies are off the table as far as I am concerned.

LTCM came into existence less than a decade after Black Monday, so everyone connected with it had certainly gone through the Black Monday experience, although maybe none of them had been inconvenienced by being unable to trade. It is possible to construct strategies that are "perfectly hedged"—that even though it is impossible to trade, your position risk is limited. If you have spent $1,000 on an option to buy 100 barrels of oil at $150 on the futures market for delivery in January 2010 and have received $600 by selling an option to buy 100 barrels of oil at $160 for the same delivery, your risk is limited—if oil goes to $300 a barrel (God help us), you will make $600. You have spent a net of $400 buying and selling the two options, and with oil at

$300 a barrel you will exercise your option, purchasing 100 barrels of oil for $15,000. The person to whom you sold the option will likewise exercise his option, purchasing those 100 barrels of oil from you for $16,000. You net $16,000 − $15,000 − $1,000 + $600 = $600.

If oil goes to $20 a barrel (saints be praised), you will lose a net of $400, the difference between the cost of the option you bought and the income from the option you sold, because neither you nor the person to whom you sold the option will exercise it. What's the point of exercising an option to buy oil at $150 a barrel when you can get it in the open market for $20 a barrel? Your risk and gain are limited when you have bought and sold the same option at different prices. At any rate, it is hard to believe that the principals of LTCM, the most market-savvy traders and theoreticians on the planet, failed to include in their calculations the results of being unable to trade. The fact that all of these brilliant and sophisticated individuals did not do the appropriate expected-value calculation points out yet again one of the great recurrent conceptual mistakes: low probability means low probability. It does *not* mean zero probability.

During the last two decades, there have been a number of high-profile investment disasters. Some, but not all, arise from the same type of scenario that caused the collapse of LTCM. One of the most highly publicized of these was the 1994 bankruptcy of California's Orange County, which has one of the highest per-capita incomes in the state. This particular fiasco was the result of Orange County's treasurer Bob Citron investing in risky interest rate securities. In retrospect, this catastrophe stemmed not from ignorance of expected value, but from attention to it. In the late 1970s, California passed Proposition 13, which limited the revenues that cities and counties could derive from property taxes. This revenue shortfall led to Citron's realizing that the expected value of secure investments was insufficient to the needs of the county. Citron saw the only alternative as riskier investments, which backfired—to the tune of $1.7 billion.

Citron was not the first "rogue trader" to take advantage of the ability to invest huge sums without adequate controls. Risky investments in futures markets by Nick Leeson brought down the centuries-old Barings Bank in 1991. Early in 2008, the venerable Societe Generale was hit with a $7.2 billion loss (and you thought losses at that level could occur only from incompetent or corrupt political machinations) in unauthorized trading due to the actions of Jerome Kerviel. Unlike Citron, who was in charge of the Orange County investment pool, Leeson and Kerviel were nowhere near the top of the investment food chain at their respective institutions. Because situations like this have happened, the probability of their happening is not zero. As a result, when a financial institution undertakes investment strategies, it is either necessary to take action to reduce the probability of such nefarious actions to zero or incorporate this into the expected-value calculation of employing such strategies.

An Expected-Value Analysis of Hurricane Katrina

An expected-value analysis of the *Challenger* disaster would almost certainly have saved the lives of seven astronauts and one space shuttle, but an expected-value analysis of a project to strengthen the levees around New Orleans, performed after the 1986 report of the Army Corps of Engineers, could have resulted in the savings of hundreds of lives—and a city.

I have not seen an estimate of the cost (in 1986 dollars) of a project to strengthen the levees to a point where they could have withstood Hurricane Katrina. Nor have I seen a 1986 estimate of the probability of a storm such as Hurricane Katrina. What *is* publicly available, though, is the human cost of Hurricane Katrina (more than 1,800 lives lost, tens of thousands of lives interrupted or ruined), and the property damage (more than $80 billion).

Hurricane Katrina was a perfect storm, but it was not unprecedented. It was the sixth largest Atlantic storm ever recorded and the third strongest to hit the United States. There is a database of hurricanes (HURDAT) going back to 1851 from which the probability of a storm such as Hurricane Katrina could have been estimated.[5] If the probability of such a storm impacting New Orleans was 1/1,000, the damage would have an expected value in the range of $100 million; if the probability was as high as 1/100, the damage would have an expected value of approximately $1 billion. If nothing else, cities that are vulnerable to such storms should at least have some sort of estimate done as to the cost of protecting against the unthinkable.

How Math Can Help

Three disasters. The first, *Challenger*, can and should have been avoided. There is simply no excuse for not doing the math when there is absolutely no downside to doing so.

The second, LTCM, might have been avoided, but the combination of circumstances that caused it to occur was so unusual that an expected-value analysis might not have raised any red flags. The process of doing the analysis, however, might have resulted in the realization that potentially unlimited risk was being assumed—but those in charge might have ignored it. Many of humanity's greatest advances have been accompanied by potentially unlimited risk; Columbus's voyage of discovery is an obvious example.

The third disaster, Hurricane Katrina, should serve as the ultimate example of how important it is to do the math. If we don't know what the price of a disaster will be, and expected value gives us the long-term average price of a disaster, how can we possibly decide whether it is worth our while to try to guard against it?

It would probably surprise most people to know that the city currently judged most at risk of a levee breach is located neither in a hurricane zone nor on an ocean. It is, in fact, in the heart of central California. Sacramento lies on the Sacramento River just below the juncture of the American River. Filmgoers may recall scenes in *Indiana Jones and the Temple of Doom* that were filmed on cliffs ostensibly near a river in India; it was actually the American River. Hurricane Katrina prompted an evaluation that led to people realizing the acute danger of a levee breach that could have catastrophic consequences for Sacramento. This prompted a plan for Sacramento to complete levee reinforcement by 2010. As of January 2008, however, levee reinforcement ran head-on into the most rapidly developing area of Sacramento, and as of this writing, public safety and economic expansion have not yet resolved their conflict.

A Tale of Two Cities

I'll close this chapter with a tale of two cities from the summer and the fall of 2007. It's almost like reading the classic tale about the ant and the grasshopper, the one where the ant prepares for the winter by laying in a store of food, while the grasshopper parties like it's 1999.

The Canyon fire in Malibu started around five in the morning of October 21. Malibu is an upscale beach community with some very famous residents, many beautiful and expensive homes, and a long history of devastating fires. Los Angeles County, like the ant, had prepared for the possibility by renting super scoopers for the fire season. Super scoopers are planes with the capacity to pick up large quantities of water from nearby sources—in this case, the Pacific Ocean or even swimming pools on the Malibu estates of the rich and famous. Renting these planes isn't cheap, but from an expected-value standpoint, it's an absolute bargain, considering the frequency with which fires occur in Southern California. The Canyon fire destroyed twenty-two buildings, and three people were injured.

San Diego, however, was nowhere near as well prepared. Amazingly enough, San Diego's only aerial fire defenses consisted of a few helicopters of Vietnam War vintage. Although it is not clear how much damage could have been prevented had San Diego been better equipped to deal with the fire in its early stages, approximately 500,000 acres were burned; 1,500 buildings were destroyed; and nine lives were lost as a direct result of the fire.

The lessons from *Challenger*, LTCM, Katrina, and San Diego are all the same lesson, and it is simple: do the math, and then use the results to plan intelligently. Math could have helped prevent two of these disasters and limited the damage from the others, but math doesn't sit there and get done on its own—somebody has to do it, and once it is done, we have to make the best of what we learn from doing it.

13

How Math Can Improve Society

How much is a human life worth in dollars?

• • •

When should legal cases be settled out of court?

• • •

At what point does military spending become unnecessary?

ecause this book centers on arithmetic, it probably isn't surprising that many of the areas in which we have applied arithmetical techniques deal with money. Money is the means by which we conduct commerce, and arithmetic is how we keep score in financial dealings.

According to the great Russian novelist Leo Tolstoy, governments are associations of men who do violence to the rest of us. Tolstoy was also an anarchist, but he certainly summed up the feelings of many who are irritated by the governments that run the cities, the states, and the countries of which we are a part. Conservatives strike a responsive chord when they talk about government being the problem, rather than the solution.

Certainly, governments sometimes do substantial harm and often could do more good than they actually do. Mathematics plays a role in both types of situations, and this chapter discusses some of the good that mathematics could do that it doesn't, and some of the harm that governments do to their citizens that has a mathematical component. It probably won't come as much of a surprise that a lot of this has to do with government's handling—and mishandling—of financial situations.

The Firefighter and the Dog Food

If you live outside Los Angeles, you've probably never heard the name Tennie Pierce. Tennie served the Los Angeles community for nearly two decades as a firefighter. Pierce stands a rugged six feet five inches and goes by the nickname of "Big Dog." In the volleyball matches that firefighters often have to pass time and stay in shape between fighting fires, Pierce often declaimed that the other players should feed the Big Dog, and he spiked the ball away for a winner.

The firefighter culture in some ways resembles a college fraternity, filled with hazing and practical jokes. Pierce participated in many of these, often on the side of the jokers. One day, however, he was served a plate of spaghetti at the firehouse. As Pierce ate the spaghetti, some of the other firefighters snickered, knowing that their colleagues had spiked Pierce's spaghetti with dog food. They had, indeed, fed the Big Dog—with dog food.

One would think that in the normal course of events in the firehouse, such an incident would quickly be forgotten. In fact, originally it was—Pierce did not seem to make much of a fuss over it.

Did I forget to mention that Pierce is black? I also forgot to mention that although most readers are undoubtedly aware of it, we live in an environment where harassment lawsuits have become a closet industry. Pierce sued the city of Los Angeles for racial harassment and intentional and negligent infliction of emotional distress on the part of his fellow firefighters. The case was to be argued by a high-powered attorney. Fearing a possible adverse judgment from a "downtown jury"—code for a predominantly black jury of the type that acquitted O.J. Simpson in his famous double-murder trial—the city attorney recommended settling the lawsuit for $2.7 million. The city council voted 11 to 1 to support such a settlement.

A major public outcry followed, stimulated by the hosts of a popular drive-time radio talk show airing opinions that such a settlement was lunacy. Responding to a sense of public outrage, the Los Angeles mayor vetoed the settlement. As a result, the lawsuit moved closer to a trial. The city attorney hired an outside law firm as consultants to conduct focus groups and mock trials in an attempt to determine the most probable outcome of the trial. The recommendation was that the city should settle, because an award in the neighborhood of $7 million was considered a possibility.

The ultimate resolution came when the mayor announced that the suit had been settled out of court, with Pierce to receive $1.5 million. Touted as a victory by all concerned, it actually represented a loss to the taxpayers—of $4.4 million. Not included in the $1.5 million settlement were $1.3 million in expenses paid to the outside law firm (perhaps $1.3 million in billings might be a more accurate description) and an additional $1.6 million paid in the settlement of a lawsuit by Pierce's superiors, who claimed that as a result of the firestorm surrounding the original case, they had been unfairly suspended.[1]

Two Conclusions

To be fair, I should say at the outset that I know very little about the legal system. I have been an alternate juror twice. It's the worst of both worlds: you have to pay attention, but you don't get to vote. There are two conclusions, however, that seem to me fairly straightforward: one arithmetical, one logical.

The arithmetical conclusion is that paying for an outside law firm is possibly even less of a good bet than buying a service contract on a refrigerator. Much of the time, as in this case, these lawyers will reach the identical conclusion that the city district attorney did, that you should settle. When they reach the opposite conclusion, whose word are you going to take—theirs or that of the city district attorney, who at least can be expected to know the territory? In addition, the cost of the outside law firm in this instance was almost 50 percent of the proposed settlement. How can one even consider buying a service contract for 50 percent of the cost of the merchandise?

The logical conclusion, which also involves some arithmetic, is that the city should fight such cases tooth and nail. It's a little harder to figure the expected value of fighting here, because even if there is an adverse decision, it's not clear what amount the jury will propose for the settlement. Although it's hard to believe that a few bites of dog food could inflict $2.7 million worth of emotional distress, to believe that it can inflict $7 million worth of emotional distress is almost beyond belief.

Arithmetic in the Courtroom

How does a jury arrive at the amount of a settlement in a civil case? I have some firsthand knowledge of this from my experience as an alternate juror on a civil case, which the plaintiff won. All of the jurors, excluding the alternates, were asked to decide the amount of the settlement. The judge then gave the jury an instruction that left me absolutely baffled, and I have a very high threshold of baffle. In deciding the amount of the settlement, jurors were forbidden to use any arithmetical process, such as computing an average of the amounts suggested by the individual jurors. It was all I could do to refrain from saying that *any* attempt to determine the amount of the settlement, because it involves numbers, constituted an arithmetical process. Had it been a department meeting, I would have opened my mouth, but I didn't fancy a stay in the graybar hotel for contempt of court. Even if one person were to suggest a settlement, and someone else would say, "That seems high to me," the second person has used the arithmetic process of comparison. If anyone who has some connection with the judicial system reads this, let me make a suggestion for determining the amount of a settlement: have the jurors discuss the issue for a set period of time, and then have every juror name a figure that he or she considers to be a fair settlement. Use "gymnastics scoring" to compute the amount of the award: throw out the two high and the two low numbers, and take the average of the remaining eight.

Returning now to an estimate of the expected value of fighting the case, it would help to have a "track record" of similar lawsuits from which to estimate both the probability of an unfavorable verdict and the likely cost of such. Yet in estimating the probability of an unfavorable verdict, it should be mentioned that the decision must be unanimous: if just 1 juror out of 12 believes that the lawsuit is frivolous or otherwise unwarranted, the city wins its case. For reference, if 95 percent of the population believes that the firefighter deserves to win the case and a jury of twelve is randomly selected, the probability of at least one juror believing that the firefighter should not win the case is about 46 percent. If only 90 percent of the population believes that the firefighter deserves to win the case and a jury of twelve is randomly selected, the probability of at least one juror believing that the firefighter should not win the case rises to 72 percent. If one-sixth of the population believes that the firefighter should not win the case, the probability that the jury will contain at least one such person is about 90 percent. The expected value of contesting the case certainly seems to be considerably below the originally proposed settlement of $2.7 million.

Moreover, logic would dictate that if cases like this are routinely settled (as they appear to be in Los Angeles), the ease with which settlements are obtained would tend to produce more and more plaintiffs seeking such settlements with ever more marginal excuses. I once jokingly suggested to a female colleague with a sense of humor that we could augment our pensions via the following strategy: the first one to retire would be sued by the other for sexual harassment, and both parties would split the settlement. Of course, such a suggestion was facetious in our case, but considering the ease with which "lottery ticket" settlements are obtained, I wouldn't be surprised to read of such a situation sometime in the future.

Bureaucratic Depreciation and the Devaluation of Human Life

Unfortunately, any mathematical tool is a two-edged sword. It can be used both to improve your life and to devalue it. Such was the case in July 2008, when an Environmental Protection Agency office lowered its valuation of human life from $8.04 million to $7.22 million.[2]

This figure is used as the estimate of the value of a typical American life when computing whether a particular measure is cost-effective. Of course, this is the same type of computation that a pharmaceutical company makes when deciding whether to pursue research to find a cure for a particular disease. If only a hundred people in the United States contract such a disease and it would cost an estimated $200 million to undertake a successful program to cure that disease, the cost is therefore $2 million per person. Unless Bill Gates contracts the disease, it's probably going to be fairly hard to submit an insurance claim for $2 million—and this would only enable the company to break even.

When judging the value of a piece of environmental legislation, the EPA first estimates how many lives will be saved.

Legislation to reduce air pollution, for example, would reduce the number of deaths due to asthma. Suppose that a program to reduce air pollution is estimated to save 4,000 lives, and the available money to fund the program is $30 billion. If a human life is valued at $8.04 million, the value of the 4,000 lives is more than $32 billion, and the program is worthwhile, at least from the standpoint of simple economics. If, however, a human life is valued at $7.22 million, the value of the 4,000 lives is less than $29 billion. As a result, the program doesn't make economic sense.

Devaluing human life, at least financially, therefore has the effect of reducing the willingness of the government to spend money protecting it. Nonetheless, you can be reasonably certain that the $30 billion available will be spent somewhere else. It might actually be that the EPA finds a more cost-effective way to spend the money. Alternatively, perhaps the money will be diverted from the EPA to be spent elsewhere. Let's see where it might be spent by looking at one of the most popular ways that the government has spent money in the past.

The aircraft carrier USS *Ronald Reagan* was commissioned in July 2003 and was built at a cost of $5 billion.[3] It is truly a magnificent piece of hardware and is manned by a crew of more than 5,500 dedicated men and women. Considering the firepower and sophistication of both the ship and the aircraft it carries, it could probably have won many of the naval battles of World War II on its own.

But we're not fighting World War II anymore. We already had a number of extremely effective aircraft carriers prior to the construction of the USS *Ronald Reagan*, and it is hard to see how the *Reagan*'s existence adds significantly to keeping the United States safe. Just as a computation should be made to check on the value of the money spent on environmental legislation by assessing how many lives the program will save, one should also make some sort of calculation as to how many *additional* American lives will be saved by the USS *Ronald Reagan*. My guess is that the most probable number is zero—but it would have to save an extra 700 or so lives (accepting the EPA' s current valuation of $7.22 million per person) in order to justify its existence economically. That's not even including operating costs, which come to $2,500,000 for every day that the *Reagan* is at sea and $250,000 for every day that it is in port. If the *Reagan* is at sea half the year (and if it isn't, why did we even build it if all we are going to do is keep it in port?), that comes to approximately half a billion dollars a year. That's 70 extra people a year it should be saving. It seems only fair that if the government is going to veto programs according to the "cost of human life" criterion, it should not embark on other programs that also fail to meet this criterion. The only naval threat on the current horizon seems to be from Somali pirates—and as of this writing, the body count from the actions of Somali pirates favors us by 3 to 0.

The truly ironic thing about the construction of the USS *Ronald Reagan* is that Reagan himself might have vetoed it had he crunched the numbers. Say what you will about the Gipper, he was a staunch proponent of smaller government and cost-effective government, and it's really hard to see the USS *Ronald Reagan* as an example of cost-effective government.

The United States has the most powerful army in the world, by an order of magnitude, on which it spends over $300 billion annually. Is this really necessary? It was certainly necessary to spend whatever it took to win World War II, which was one of those situations that we simply could not afford to lose. It might even have been a good idea to ramp up our military during the cold war. Yet even though there are disquieting rumblings out of Russia and China, the two main announced threats are Iran and North Korea, either of which would probably lose any military conflict with the United States in short order. Not only that, inordinate sums are being spent trying to prevent terrorists, who are mostly small groups of individuals, from performing destructive acts. I, and many others, have doubts about how cost-effective this is.

Bracket Creep and Fiscal Drag

The best taxes, at least from the standpoint of the agency collecting the taxes, are stealth taxes: taxes that the individual being taxed may not even be aware he is paying. Almost all income taxes, whether collected by the Internal Revenue Service or by the state agencies for states that assess income tax, are based on brackets: an income range in which the tax is a specific percentage. When a person's income jumps from one bracket to the next, the additional income is taxed at a higher rate. For example, dollars earned between $40,000 and $50,000 may be taxed at 10%, but dollars earned between $50,001 and $75,000 may be taxed at 12%.

Many employees—and all Social Security recipients—receive COLA raises, which are cost of living adjustments designed to compensate for inflation. Suppose that in the previous example, a person is making a salary of $49,000 and receives a 3% COLA raise to compensate for inflation. This raises the salary to $50,470, putting the employee in the next tax bracket. If the brackets—the numbers that define the ranges in which the tax percentage remains constant—are not raised to compensate for inflation, the employee finds that $470 is taxed at 12% rather than at 10%. This phenomenon is known as bracket creep.

This example may not seem like a lot of money, especially for one individual. California, however, which is currently experiencing a major fiscal crisis, is considering leaving its income brackets unchanged, and this is expected to add more than $1 billion to the state's coffers next year.

Bracket creep is an extreme case of fiscal drag, where the government increases the tax brackets (thus avoiding the accusation of bracket creep) to compensate for inflation but does so at a lesser rate than inflation. This can result in a higher effective income tax rate, even though the nominal income tax rate doesn't change.

Assume that the tax rate is 20% on all earnings above $10,000. An individual earning $50,000 thus pays 20% of

$50,000 − $10,000 = $40,000, or $8,000. Thus, $8,000 is 16% of $50,000, and this is the effective tax rate that the individual is paying.

Suppose that this individual receives a COLA of 5%, and the government raises the brackets by 2%. The individual is still paying 20% on income above the bracket minimum, which is now $10,200 ($10,000 plus 2% of $10,000). Because 5% of $50,000 is $2,500, the individual's income is now $52,500. He pays a tax of 20% on $42,300, the difference between his income of $52,500 and the bracket minimum of $10,200. This is 20% of $42,300, or $8,460. A tax of $8,460 on an income of $52,500 is an effective tax rate of 16.11%. Not much—for this individual—but it adds up (more than enough for bureaucrats to vote themselves pay raises beyond the inflation level), and nobody notices, except the eagle-eyed.

A variant of bracket creep can also have an effect on your pocketbook, although it is not necessarily clear that this particular maneuver is done with that intent. Various medical associations issue guideline numbers for an assortment of tests. Currently, a level of total cholesterol over 200 is considered to be unhealthy, as is a glucose total in excess of 100. Both of these numbers, delineating the lower end of the "unhealthy" bracket, were recently lowered. Prior to this, an individual with a total cholesterol reading of 210 was deemed to be healthy; after bracket creep set in, such an individual might be a potential candidate for cholesterol-lowering drugs. I'm willing to give doctors the benefit of the doubt and assume that they lowered the cholesterol bracket with good intent. I'm even willing to go a little further and assume that the vast majority of doctors who have a patient whose total cholesterol reading is 210 will advise changes in diet and exercise before writing out prescriptions for Lipitor. Nonetheless, it's a fair bet that more prescriptions for Lipitor are written out with a bracket whose lower limit is 200 than with one whose lower limit is 220.

14

How Math Can Save the World

Do extraterrestrial aliens exist?

• • •

How can we prevent nuclear war and a major
asteroid impact?

• • •

When is the world going to end?

What, exactly, is mathematics? Dictionary.com (used by those of us who spend more time online than at the library) defines it as "the systematic treatment of magnitude, relationships between figures and forms, and relations between quantities expressed symbolically." That's what mathematics *is*, but this book is much more concerned with what you can *do* with it, which is why mathematics goes far beyond the dictionary.com definition. Probably the most dramatic thing you can do with it is save the world.

Maybe that's a little bit of hyperbole, but those of us who work with mathematics are continually surprised by its ability to enable us to evaluate, to predict, and to plan. Of course, mathematics by itself is incapable of saving the world, but we can look at some of the possible threats to the world, predict how likely they are, evaluate whether we can do anything about them, and plan how best to use our resources.

I don't think that many readers of this book spend sleepless nights worrying about apocalyptic predictions of the End of Days or the threat of takeover by aliens, but both of these afford interesting examples of mathematics in action. End of Days predictions seem (so far) to have a pretty low batting average (remember the great Y2K catastrophe and the Jupiter Effect?) but sufficiently high entertainment value that they spawn a lot of literature and TV specials. My favorite End of Days scenario is the Tower of Hanoi problem, which is worth a little time and effort because (1) it's mathematics, and (2) it's cute, and there's an echo of it in Arthur C. Clarke's classic science-fiction short story "The Nine Billion Names of God," which has one of the best last lines of any story I've ever read.[1] I give myself points for recognizing a great story when I read one; when I researched the source to footnote this story, I discovered that it had won the 2004 retrospective Hugo award for the best science fiction story of 1953.

The Tower of Hanoi

I first saw the following passage in a book I read as a child; I'm not enough of a historian to chase it back to its roots, but I will give a Web reference (again, due to spending more time on the Web than at the library).

In the great temple at Benares, says he, beneath the dome which marks the centre of the world, rests a brass plate in which are fixed three diamond needles, each a cubit high and as thick as the body of a bee. On one of these needles, at the creation, God placed sixty-four discs of pure gold, the largest disc resting on the brass plate, and the others getting smaller and smaller up to the top one. This is the Tower of Bramah. Day and night unceasingly the priests transfer the discs from one diamond needle to another according to the fixed and immutable laws of Bramah, which require that the priest on duty must not move more than one disc at a time and that he must place this disc on a needle so that there is no smaller disc below it. When the sixty-four discs shall have been thus transferred from the needle on which at the creation God placed them to one of the other needles, tower, temple, and Brahmins alike will crumble into dust, and with a thunderclap the world will vanish.[2]

I think this is probably where Clarke got the idea for "The Nine Billion Names of God." At any rate, it's fairly easy to see how it works. Let 1 be the number of the disk of the smallest diameter and 64 the number of the disk with the largest diameter. We'll use a diagram to keep track of where the disks are on the various needles.

	Needle A	Needle B	Needle C	Disks Moved
Start	64 up to 1			

For the first move, we move disk 1 from Needle A to Needle B.

	Needle A	Needle B	Needle C	Disks Moved
1st move	64 up to 2	1		1

By this, I mean that we've gotten a small stack (1 disk) transferred to another needle with the largest on the bottom.

	Needle A	Needle B	Needle C	Disks Moved
2nd move	64 up to 3	1	2	
3rd move	64 up to 3		2 up to 1	2

Now Needle C contains a stack of 2 disks (the smallest and the next smallest), with the largest on the bottom.

	Needle A	Needle B	Needle C	Disks Moved
4th move	64 up to 4	3	2 up to 1	
5th move	64 up to 4, 1	3	2	
6th move	64 up to 4, 1	3 up to 2		
7th move	64 up to 4	3 up to 1		3

And now Needle B contains a stack of the three smallest disks, with the largest on the bottom and the smallest on top.

Show the pattern 1,3,7 to a mathematician, and he or she will notice that each number is 1 less than the next power of 2:1 is 1 less than 2, 3 is 1 less than 4, and 7 is 1 less than 8. He or she will then conjecture that it takes $2^{64} - 1$ moves to transfer 64 disks.

It's not hard to see why this is true. Suppose you've already created a stack on another needle with disks 1 through 26 in the appropriate order. You now move the 27th disk to a free needle and recreate the entire sequence of moves you've just gone through to put disks 1 through 26 on top of disk 27. Therefore, however many moves it took to move a stack of 26 disks, it will take twice that number plus one to move a stack of 27 disks. That's precisely the rule that generates the sequence 1,3,7,15,31, ... Each number is twice the preceding number plus one, and each number is also one less than the appropriate power of 2.

Now that we know how many moves it's going to take, let's estimate how long it will be until the end of the world. If the monks are reasonably agile, they might be able to transfer 1 disk every three seconds. It would then take $3 \times (2^{64} - 1)$ seconds. This is about 1,750,000,000,000 years—and considering that the universe has been around only 14 billion years or so, it's my preferred end-of-the-world scenario. I'm not going to be around for the remaining 1,736 billion years (unless cloning becomes perfected and affordable for math professors), but it's been an interesting show so far and I'd prefer not to see it close early. Notice that the previous passage refers to the end of the worlds—possibly meaning the end of the universe, but the time remaining for Earth is considerably less, because the Sun is scheduled to expand to a red giant and consume Earth in a few billion years. Maybe when the Sun starts to expand, the monks can move the needles and the disks to a galaxy far, far away and continue the project.

The Drake Equation

I'm not sure whether H. G. Wells was the first to envision an invasion by a sentient alien race bent on world domination, but his classic sci-fi story *War of the Worlds* has been made into two successful movies and has spawned countless imitators. Half a century ago, a group of scientists envisioned kinder and gentler aliens than the ones postulated by Wells. These scientists met in Green Bank, West Virginia, in order to found an endeavor now known as SETI, the Search for Extra-Terrestrial Intelligence. One of the participants was Frank Drake, who proposed an expected-value calculation to estimate N, the number of civilizations in the galaxy with whom communication might be possible. This expected-value calculation was presented in the following formula, which is now known as the Drake equation:[3]

$$N = R^* \times f_p \times n_e \times f_l \times f_i \times f_c \times L$$

Each of the seven factors on the right side of the equation is basically a guess. What follows is a definition of those factors, Drake's original estimates of the numbers, and a little about the current thinking concerning the values of those numbers. After I present this information, Drake's estimate of N will be computed, along with some discussion as to how that number might change due to more recent information.

R^* represents the rate of star formation in the galaxy: how many new stars are created each year. Drake estimated this number as 10; better technology over the last fifty years has resulted in NASA estimating this number as 7.

f_p is the fraction of those stars that have planets. Drake, with nothing to guide him other than the solar system, estimated this number as 0.5. Planet hunting has now evolved into a fine art. More than 300 planets are known to exist outside the solar system, and our current technology enables us to find only really big planets. Drake's estimate was certainly a guess—for the time being, let's keep the number as 0.5, with the understanding that it could definitely be higher. I don't know whether *any* stars have been found that are known to have *no* planets.

n_e is the average number of habitable planets per star that has planets. Drake estimated this as 2, but the consensus today is that this number is probably much smaller. The habitable zone, where the temperature of the planet is neither too hot nor too cold to support life, is generally fairly narrow. In addition, the parent stars must have a sufficiently long period of stability and must supply sufficient heavy elements to support life. There's no consensus on this that I could find, but relatively few planets have been found in habitable zones.

f_l is the fraction of habitable planets on which life develops. Drake used 1 for his estimate, and recent arguments based on the length of time it took life to evolve on Earth have concluded that this fraction is greater than 0.13. The question boils down to this: given the right conditions, how inevitable is life?

f_i is the fraction of planets with life that go on to develop intelligent life. Drake estimated this as 0.01. Nobody has a clue, and this guess is probably as good as any.

f_c is the fraction of planets with intelligent life whose species evolve the ability to communicate with others and are willing to do so. Again, Drake guessed 0.01. Earth had intelligent life for hundreds of millions of years before a combination of events triggered the ascent of mammals, the eventual emergence of man, and the development of a technological civilization, so to me this number seems high, but who knows?

L is the expected lifetime of such a civilization for the period that it can communicate across interstellar space. Drake guessed that this was 10,000 years. Our civilization has had this ability for less than 100 years, but for all we know, once the growth problems of a civilization have been surmounted, such a civilization might last for millions of years.

Drake computed $N = 10 \times 0.5 \times 2 \times 1 \times 0.01 \times 0.01 \times 10,000 = 10$. The last three numbers in particular are huge guesses; if intelligent civilizations are long-lived, this number could increase by a factor of 1,000. If intelligence is rare, however, and the evolution of a technological civilization is equally rare, this number could sink to well below 1. So, who knows?

On the other hand, we're not confronted with the problem of determining how many civilizations are out there looking to talk to us; we're wondering about how likely the scenarios depicted in *The War of the Worlds* or *Independence Day* actually are. So let's start from the Drake equation and do a little expected-value calculation of our own. Let H be the number of hostile civilizations able to get here and desirous of annihilating us for whatever reason. Then

$$H = N \times f_a \times f_g \times f_h.$$

N is the number of civilizations able to communicate with us, as determined by the Drake equation; f_a represents the fraction of those civilizations that are actually able to get here from wherever they are. There aren't any in the solar system, and, from all we know of physics, getting here is a whole lot more difficult than *Star Trek* would have us believe.

Next, f_g is the fraction of civilizations that are able to get here who will actually get here. First of all, it's a big galaxy, and we're off in the 'burbs. A civilization that can get here might just decide that prospecting is better in a more densely crowded section of the galaxy. Compounding this is the economics of getting here; this could be too expensive in terms of time, effort, and resources for a civilization to bother. Considering the fact that we can accomplish a lot simply by doing things on the Internet, an advanced civilization might just decide to chat with us, if indeed these aliens notice us at all.

Finally, f_h is the fraction of those civilizations that actually get here who do so with hostile intent—or with hostile inadvertence, much as we accidentally step on ants without even noticing it. One thing is for certain; if they can get here, they can certainly overcome whatever resistance we try to put up, but will they even care? In the same decade that Arthur C. Clarke wrote "The Nine Billion Names of God," he also wrote the novel *Childhood's End*, about the arrival of an extraterrestrial civilization whose mission was to bring about the next step in the evolution of man.

All things considered, given even the highest possible value of N that optimistic estimates might allow, I'm convinced that the small values of the other factors make it extremely unlikely that we'll ever have to worry about hostile invasions from outer space. I'm *much* more worried about the remaining two scenarios in this chapter—and for what I think are good reasons. They both present situations in which expected-value calculations *can* provide a clear guide to the steps that should be taken to save the world.

February 5, 1958

As a native New Yorker and an adopted Angeleno, I haven't put the southeast portion of the United States on my must-visit list—and after seeing a TV program on the events of February 5, 1958, I'm certainly not planning to do so. On that evening, a B-47 Stratojet bomber had one of its wings accidentally clipped in a mid-air collision with an F-86 Saberjet. The pilot of the wounded B-47, Major Howard Richardson, performed a task every bit as impressive as the one Captain Chesley Sullenberger completed half a century later: he managed to bring his badly crippled plane back without loss of life. To do so, however, he had to jettison his payload: a nearly 4-ton Mark 15 hydrogen bomb, with an explosive power of 1.5 megatons of TNT. The bomb buried itself in the muck at the bottom of the ocean off Tybee Island—where, to the best of everyone's knowledge, it still remains. The U.S. Air Force hasn't found it, salvage operations haven't found it—at least, the salvage operations that we know of haven't found it—and it's still sitting there, half a century later.[4]

If the bomb goes off, it won't end life on this planet. If a bunch of them go off, however, they could. And if a lot of them go off, they almost certainly will. Viewed in terms of expected value, the number of lives lost is related to the probability of a bomb going off in a populated area and the explosive power of the bomb. For the most part, the superpowers that are the source of most of the world's supply of atomic and hydrogen bombs have done a remarkable job; there has been no hostile use of one of these weapons since August 9, 1945, the day the second atomic bomb was detonated over Nagasaki. The fall of the Soviet Union led to concerns that part of its nuclear and thermonuclear arsenal might fall into the hands of those who might be impelled to use them, no matter what the cost to themselves, but to date no catastrophic events have occurred.

I am not privy to the bookkeeping on the world's nuclear arsenal. Most of it is accounted for, and *Dr. Strangelove* notwithstanding, the probability of a rogue U.S. military officer being able to launch or drop a nuclear weapon seems extremely small. We can only hope the same is true for the nuclear weapons produced by the other nuclear powers. Yet there is a major step that we and the other nuclear powers can take to reduce the possibility that the planet will end in a thermonuclear holocaust. We can reduce the explosive power of the nuclear weapons and, by so doing, significantly change the expected value (in terms of human life) of their existence.

Nuclear weapons exist, and while there are those who feel it would be nice if these swords were beaten into plowshares, it's almost certainly not going to happen. Yet what is the real need for thermonuclear weapons? If a nation needs nuclear weapons as a deterrent, A-bombs do just as good a job as H-bombs, and with a small fraction of the equivalent tonnage of TNT. The bombs that exploded over Hiroshima and Nagasaki are rated at 20,000 tons of TNT, and one merely has to look at the postapocalyptic pictures of those cities to realize that the same bomb detonated over *any* major city in the world would wreak untold havoc. The largest H-bomb ever was rated at 50,000,000 tons of TNT. If there is such a thing as an ultimate deterrent, an atomic bomb is it. If something insane happens and a few of them go off, the planet can survive. Whether the planet can survive a few large H-bombs is a matter of some scientific debate. No nation reduces its deterrent by reducing the size of the payload in its nuclear arsenal, but every nation benefits if they all adopt such a strategy.

A Visit to the Yucatan Peninsula

It was almost certainly a warm day, because the Cretaceous Period was marked by substantial warmth. Some estimates are that during the warmest period, sea-surface temperatures were well over 100 degrees, warmer than many heated pools. At any rate, the profusion of life during that period indicates that while global warming may drastically alter life on Earth, it will not eradicate it. Yet an event that occurred on that almost certainly warm day nearly did.

No human eyes were around to witness the cataclysmic events of that day, but it must have been an amazing sight. A fireball streaked across the sky, and a huge meteor struck the Yucatan Peninsula in Mexico, leaving a massive hole known as the Chicxulub crater (from the town located near its center) more than 110 miles in diameter.[5] The impact is estimated to have been equivalent to 100 trillion tons of TNT; for contrast, the largest H-bomb ever detonated was approximately 50 megatons, so the meteor impact was equivalent to 2 million such bombs, all going off at the same time in the same place. To say that this was an Earth-shaking event is an incredible understatement. It wiped out nearly 70 percent of the species on the planet, ending the 160-million-year reign of the dinosaurs. Bad for the dinosaurs—but good for us. Mammals, which had only managed to eke out a foothold during the era of the dinosaurs, quickly filled many of the ecological niches left vacant by the aftermath of the meteor impact. Sixty-five million years later, here we are.

The meteors are here, too. One detonated in an air burst over Tunguska in 1908, with an estimated blast energy of between 5 million and 30 million tons of TNT. It occurred in so remote an area that the loss of life was minimal, but the next one may hit New York or Tokyo.

Even worse, the next one may be larger—much larger. The Tunguska meteor is estimated to have been a few tens of meters in diameter. According to a paper published by the Jet Propulsion Laboratory, a kilometer-sized meteor impacts Earth on the average of once every million years, an event that would threaten human existence. A Chicxulub-scale event happens every 50 to 100 million years and would almost certainly clear the way for whatever species takes over from us—if any are left to do it.

We are the only species in the history of the Earth—and possibly in the history of the universe—to be able to avert such a catastrophe.

99942 Apophis

For a short period in 2004, it was felt that the probability of an asteroid of significant size hitting Earth was almost exactly equal to the probability of rolling snake eyes (double aces) with a pair of dice.

Apophis (named for an alien who tried to destroy Earth in the TV series *Stargate SG-1*) is a near-Earth asteroid, more than a thousand feet long, whose orbit will bring it close to Earth in 2029—much too close for comfort. The current estimate is that it will pass *below* the level of the geosynchronous satellites we currently have in orbit, but it will, thankfully, miss Earth. If it doesn't, we can expect an impact that will release more than five times the energy of the 1883 explosion of the island of Krakatoa, the most powerful event to occur on Earth in recorded history. In addition to the usual devastation due to a volcanic explosion, Krakatoa altered Earth's climate for nearly five years. If Apophis hits, we can expect a meltdown of almost apocalyptic proportions—but it appears likely that it will not hit in 2029. Although at one time projections gave it approximately a 2.7 percent probability of impact—about the probability of rolling snake eyes (talk about "crapping out"!), subsequent refined measurements eliminated the possibility of its hitting Earth in 2029. Like the Terminator, however, it will be back to try again—in 2036 and 2037. Its close pass to Earth in 2029 will alter its orbit, and precise estimates of the 2036 and 2037 impact probabilities will not be available until accurate radar measurements can be taken in 2013.[6]

NASA takes these things very seriously, as should we. Apophis was discovered in June 2004, about twenty-five years before the original estimate of possible impact. There are a lot of asteroids around, and we have the technology to find and track almost all of the dangerous ones. Even more important, we have the technological capability to avoid many of the potentially catastrophic ones—if only we find out about them soon enough and assemble the technological resources to do so.

Numerous technological options for dealing with such a situation have been discussed, from pulverizing the asteroid with thermonuclear weapons (maybe those with large warheads could be put to good use after all) to deflection strategies of all types. All of these discussions are theoretical, however, because almost no funds have been allocated for this project.

What is the expected value of a major asteroid impact? It's basically off the scale: even though the probability is low, the negative payoffs are so high as to make the expected value unacceptable. How much is it worth to us to prevent it? One would think that in an era where multitrillion-dollar budgets are being proposed to prop up a faltering economy, some spare change (maybe a few billion) could be thrown at the problem of preventing the eradication of humanity. According to a story in the August 12, 2009, online edition of *USA Today*, NASA doesn't even have the funds to detect a large percentage of these potential Earth killers, much less to develop strategies to deflect them.

John F. Kennedy's plan to put an American on the moon by the end of the 1960s did more than simply galvanize the country; it was a key factor in the aerospace boom that helped get the economy on track. Global preservation is every bit as worthwhile a goal as putting an American on the moon—and at this time it just might be the most important common goal that the people of Earth can embrace.

This book began with how an understanding of expected value can greatly improve the quality of your life. It ends, symmetrically, with a discussion of how an understanding of expected value can help save the planet. If the expected value of an event is negative—as it certainly is for both meteor impacts and thermonuclear catastrophes—there are two possible approaches suggested by expected value. We can attempt to reduce the probability of these events, or we can reduce the negative payoffs associated with them. It's hard to see how we could reduce the negative payoffs associated with a meteor impact, so we had better put all of our eggs in the "probability reduction" basket of early detection and early countermeasures. Thermonuclear catastrophes, however, are amenable to both approaches. The probability of occurrence has always been considered, which is why thermonuclear weapons are heavily guarded and are manufactured so as to reduce the probability of accidental detonation. During the height of the cold war, however, the goal was to produce terror weapons—and the higher the megatonnage, the greater the terror. Nuclear weapons are still undoubtedly needed as deterrents, but the high megatonnage that was considered a plus during the cold war is now a negative payoff, and a good place to start saving the world would be to eliminate as many of these potential planet-killers as possible. Save a few, however—just in case they're needed for Apophis.

NOTES

Preface

1. Isaac. Asimov, "The Feeling of Power" in *Worlds of Science Fiction* (New York: Quinn Publishing Co., 1958).

Introduction

1. "Could You Have Passed the 8th Grade in 1895? . . . Take a Look," Morehead State University Web site, http://people.moreheadstate .edu/fs/w.willis/eighthgrade.html.

1. The Most Valuable Chapter You Will Ever Read

1. "Why You Don't Need an Extended Warranty," www.consumerreports .org/cro/money/news/november-2006/why-you-dont-need-an-extended-warranty-11−06/overview/extended-warranty-11-06.htm.
2. See www.csulb.edu/~rmena/Discrete/Notes%20for%20Discrete .pdf. These are Professor Robert Mena's course notes for a discrete mathematics course, which includes probability theory. An easy-to-learn formal explanation can be found on page 55 of this pdf file.
3. "Game Show Problem," www.marilynvossavant.com/articles/gameshow .html. This problem provoked an absolute firestorm of controversy when Marilyn vos Savant included it in one of her columns. If you think that mathematics is so cut-and-dried that all mathematicians agree on the solution to a problem, think again—and read the e-mails that she received!
4. "College Degree Nearly Doubles Annual Earnings, Census Bureau Reports," www.census.gov/Press-Release/www/releases/archives/ education/004214.html.

2. How Math Can Help You Understand Sports Strategy

1. J. D. Williams, *The Compleat Strategyst* (New York: McGraw-Hill, 1954).
2. Ibid., p. 44. This book also shows how to handle more complicated situations in which the players have a choice of more than two strategies.
3. "Game Theory in and out of the Classroom," www.gametheory.net/students.html. For those wishing to explore current applications of game theory, this site has opportunities to do so, amusingly divided into topics for educators, students, professionals, and geeks.

3. How Math Can Help Your Love Life

1. MathProblems.info, Problem 26, http://mathproblems.info/group2.html.
2. The series is based on the equally wonderful book by Burke titled *Connections* (Boston: Little, Brown and Company, 1995).

4. How Math Can Help You Beat the Bookies

1. See www.csulb.edu/~rmena/Discrete/Notes%20for%20Discrete.pdf. These are the notes for Professor Robert Mena's course on discrete mathematics.

5. How Math Can Improve Your Grades

1. D. A. Christakis et al., "Early Television Exposure and Subsequent Attentional Problems in Children," *Pediatrics* 113, no. 4 (April 2004): 708–713.
2. "How the Test Is Scored," http://www.collegeboard.com/student/testing/sat/scores/understanding/howscored.html.
3. UCLA Law Web site, Frequently Asked Questions, www.law.ucla.edu/home/index.asp?page = 806#Undergraduate_Majors.

6. How Math Can Extend Your Life Expectancy

1. "To His Coy Mistress," www.poemofquotes.com/andrewmarvell/to-his-coy-mistress.php.
2. See Quackwatch, www.quackwatch.org/04ConsumerEducation/QA/mdcheck.html.
3. "Life Expectancy Calculations," http://space.mit.edu/home/tegmark/death.html#lifeexpec.
4. "Dennis Quaid's Twins among Three Newborns Given Drug Overdose," www.foxnews.com/story/0,2933,312357,00.html.
5. G. Marotta, "Invitation to a Tea Party," *Los Angeles Times*, April 11, 1994, p. B7.

6. "NASA's Metric Confusion Caused Mars Orbiter Loss," www.cnn .com/TECH/space/9909/30/mars.metric/.

7. How Math Can Help You Win Arguments

1. Bureau of Economic Analysis National Economic Accounts, "National Income and Product Accounts Table," www.bea.gov/national/nipaweb/ TableView.asp?SelectedTable=5&FirstYear=2008&LastYear=2009& Freq=Qtr.

8. How Math Can Make You Rich

1. "Present Value," *The Concise Encyclopedia of Economics*, www.econlib.org/ library/Enc/PresentValue.html. This Web site has concise treatments for many basic economic concepts and an assortment of references to classic texts on these subjects.
2. See FirstUSA.com for an example of one credit card's rates, www. firstusa.com/cgi-bin/webcgi/webserve.cgi?pdn=pt_chase_con_2009_ 1&card=CGT2&page_type=appterms. Credit card companies are legally obligated to define all of the conditions and terms pertaining to their credit cards. These can often be found on the back of your bill, if you receive it via regular mail. They also have posted their terms online; this is a sample.
3. "What Do Hybrid Car Batteries Really Cost?" http://money.cnn. com/2007/06/01/pf/saving/toptips/index.htm. Prices of such things as hybrid battery packs are affected by both technological advances and market conditions.

9. How Math Can Help You Crunch the Numbers

1. Spiritus-temporis Web site, "Cholera," www.spiritus-temporis.com/ john-snow-physician-/cholera.html.
2. *Journal of the Anthropological Institute*, 15 (1886): 246–263.
3. "The Limerick, a Facet of Our Culture," www.csufresno.edu/ folklore/drinkingsongs/html/books-and-manuscripts/1940s/1944- the-limerick-a-facet-of-our-culture/index.htm.
4. W. James and C. Stein, "Estimation with Quadratic Loss," *Proceedings of the 4th Berkeley Symposium on Statistics and Probability*, 1 (1961): 361–379.
5. P. Everson, "Stein's Paradox Revisited," *Chance* 20, no. 3 (2007): 49–56.
6. See "Standard Statistical Tables," http://business.statistics.sweb.cz/ normal01.jpg.
7. "Proposed Jury Instruction for Reasonable Doubt" www.state.wv.us/ WVSCA/jury/crim/reasonable.htm.

8. A serious statistician would note that this isn't precisely what a type-I error is, but this isn't a book for serious statisticians. If you would like to see the precise definition, it can be found in M. Triola, *Elementary Statistics* (Boston: Addison-Wesley, 2006), p. 398.

9. If you're interested, the computation is 500* (.12 + 1.645√((.12*.88)/500)). The progenitor formula can be found on page 408 of Triola, *Elementary Statistics*.

10. How Math Can Fix the Economy

1. Charles Mackay, *Extraordinary Popular Delusions and the Madness of Crowds* (New York: L. C. Page, 1932), p. 64.
2. J. A. Poulos, *Innumeracy* (Hill and Wang, New York, 1988).
3. "1964–Present, September 7, 1969, Senator Everett McKinley Dirksen Dies," www.senate.gov/artandhistory/history/minute/Senator_Everett_Mckinley_Dirksen_Dies.htm.
4. See "2002: Bush's speech to the White House Conference on Increasing Minority Homeownership," http://isteve.blogspot.com/2008/09/2002-bushs-speech-to-white-house.html?showComment=1222342140000.

11. Arithmetic for the Next Generation

1. See "Mathematics Content Standards for California Public Schools: Kindergarten through Grade Twelve," California Department of Education, www.cde.ca.gov/be/st/ss/documents/mathstandard.pdf.

12. How Math Can Help Avert Disasters

1. See NASA History Division, http://history.nasa.gov/sts51l.html. This is NASA's official site and contains all of the key information, reports, and links to the video, if you want to watch it.
2. For an excellent book on the LTCM fiasco, see R. Lowenstein, *When Genius Failed* (London: Fourth Estate, 2002).
3. See "New Orleans Hurricane Katrina Levee Failures," http://matdl.org/failurecases/Dam%20Cases/new_orleans_hurricane_katrina_le.htm. This site references the highlights of the major reports.
4. The Rogers Commission Report, http://science.ksc.nasa.gov/shuttle/missions/51-l/docs/rogers-commission/table-of-contents.html.
5. See the Atlantic Oceanographic and Meteorological Laboratory Web site, www.aoml.noaa.gov/hrd/hurdat/easyread-2008.html.

13. How Math Can Improve Society

1. See the *Los Angeles Times* Web site, http://articles.latimes.com/ keyword/tennie-pierce, for some of the stories surrounding the Tennie Pierce case.
2. See NaturalNews.com, www.naturalnews.com/023734.html. The value keeps declining.
3. For more information about the USS *Ronald Reagan*, see www.reagan .navy.mil/.

14. How Math Can Save the World

1. Arthur C. Clarke, *The Nine Billion Names of God: The Best Short Stories of Arthur C. Clarke* (New York: Harcourt, Brace & World, 1967).
2. See the Free Dictionary, http://encyclopedia2.thefreedictionary.com/ Tower+of+Hanoi.
3. For more on the Drake equation, see www.activemind.com/ Mysterious/Topics/SETI/drake_equation.html.
4. See "Interest in Lost H-bomb Resurfaces," www.usatoday.com/news/ nation/2004-10-19-h-bomb-search_x.htm.
5. See "Effects of the Discovery," http://palaeo.gly.bris.ac.uk/communica tion/Hanks/eff.html.
6. See "Predicting Apophis' Earth Encounters in 2029 and 2036," NASA Web site, http://neo.jpl.nasa.gov/apophis/.

INDEX